Die Grundlehren der mathematischen Wissenschaften

in Einzeldarstellungen
mit besonderer Berücksichtigung
der Anwendungsgebiete

Band 167

K. Chandrasekharan

Arithmetical Functions

Springer-Verlag Berlin Heidelberg New York 1970

Prof. Dr. K. Chandrasekharan
Eidgenössische Technische Hochschule Zürich

Geschäftsführende Herausgeber:

Prof. Dr. B. Eckmann
Eidgenössische Technische Hochschule Zürich

Prof. Dr. B. L. van der Waerden
Mathematisches Institut der Universität Zürich

ISBN 978-3-642-50028-2 ISBN 978-3-642-50026-8 (eBook)
DOI 10.1007/978-3-642-50026-8

 Library of Congress Catalog Card Number 72-102384.

Title No. 5150.

For Sarada

Preface

The plan of this book had its inception in a course of lectures on arithmetical functions given by me in the summer of 1964 at the Forschungsinstitut für Mathematik of the Swiss Federal Institute of Technology, Zürich, at the invitation of Professor Beno Eckmann. My *Introduction to Analytic Number Theory* has appeared in the meanwhile, and this book may be looked upon as a sequel. It presupposes only a modicum of acquaintance with analysis and number theory.

The arithmetical functions considered here are those associated with the distribution of prime numbers, as well as the partition function and the divisor function. Some of the problems posed by their asymptotic behaviour form the theme. They afford a glimpse of the variety of analytical methods used in the theory, and of the variety of problems that await solution.

I owe a debt of gratitude to Professor Carl Ludwig Siegel, who has read the book in manuscript and given me the benefit of his criticism. I have improved the text in several places in response to his comments. I must thank Professor Raghavan Narasimhan for many stimulating discussions, and Mr. Henri Joris for the valuable assistance he has given me in checking the manuscript and correcting the proofs.

July 1970 K. Chandrasekharan

Contents

Chapter I

The prime number theorem and Selberg's method

§ 1. Selberg's formula 1
§ 2. A variant of Selberg's formula 6
§ 3. Wirsing's inequality 12
§ 4. The prime number theorem 17
§ 5. The order of magnitude of the divisor function 19
Notes on Chapter I 21

Chapter II

The zeta-function of Riemann

§ 1. The functional equation 28
§ 2. The Riemann-von Mangoldt formula 33
§ 3. The entire function ξ 40
§ 4. Hardy's theorem 45
§ 5. Hamburger's theorem 51
Notes on Chapter II 54

Chapter III

Littlewood's theorem and Weyl's method

§ 1. Zero-free region of ζ 58
§ 2. Weyl's inequality 60
§ 3. Some results of Hardy and Littlewood and of Weyl 69
§ 4. Littlewood's theorem 73
§ 5. Applications of Littlewood's theorem 78
Notes on Chapter III 84

Chapter IV

Vinogradov's method

§ 1. A refinement of Littlewood's theorem 88
§ 2. An outline of the method 88
§ 3. Vinogradov's mean-value theorem 90
§ 4. Vinogradov's inequality 99
§ 5. Estimation of sections of $\zeta(s)$ in the critical strip 106

§ 6. Chudakov's theorem 108
§ 7. Approximation of $\pi(x)$ 110
Notes on Chapter IV 110

Chapter V

Theorems of Hoheisel and of Ingham

§ 1. The difference between consecutive primes 112
§ 2. Landau's formula for the Chebyshev function ψ 113
§ 3. Hoheisel's theorem 124
§ 4. Two auxiliary lemmas 126
§ 5. Ingham's theorem 130
§ 6. An application of Chudakov's theorem 138
Notes on Chapter V 139

Chapter VI

Dirichlet's L-functions and Siegel's theorem

§ 1. Characters and L-functions 143
§ 2. Zeros of L-functions 145
§ 3. Proper characters 146
§ 4. The functional equation of $L(s, \chi)$ 149
§ 5. Siegel's theorem 155
Notes on Chapter VI 164

Chapter VII

Theorems of Hardy-Ramanujan and of Rademacher on the partition function

§ 1. The partition function 166
§ 2. A simple case 166
§ 3. A bound for $p(n)$ 169
§ 4. A property of the generating function of $p(n)$ 170
§ 5. The Dedekind η-function 174
§ 6. The Hardy-Ramanujan formula 178
§ 7. Rademacher's identity 185
Notes on Chapter VII 191

Chapter VIII

Dirichlet's divisor problem

§ 1. The average order of the divisor function 194
§ 2. An application of Perron's formula 195
§ 3. An auxiliary function 198
§ 4. An identity involving the divisor function 200
§ 5. Voronoi's theorem 202

§ 6. A theorem of A. S. Besicovitch 204
§ 7. Theorems of Hardy and of Ingham 205
§ 8. Equiconvergence theorems of A. Zygmund 209
§ 9. The Voronoi identity 223
Notes on Chapter VIII 226

A list of books 229

Subject index 230

Chapter I

The prime number theorem and Selberg's method

§ 1. Selberg's formula. Let $\pi(x)$ denote, for any real x, the number of primes not exceeding x. The prime number theorem is the assertion that

$$\lim_{x\to\infty}\left(\frac{\pi(x)}{x/\log x}\right)=1\,. \tag{1}$$

A fundamental formula discovered by Atle Selberg has made a proof of (1) possible without the use of the properties of the zeta-function of Riemann, and without the use of the theory of functions of a complex variable. We shall prove Selberg's formula in this chapter, and indicate some of its consequences. We shall also prove an inequality due to E. Wirsing, which, when combined with a variant of Selberg's formula, gives a proof of the prime number theorem.

We recall the definitions and simple properties of a number of well-known arithmetical functions. An *arithmetical function* is any complex-valued function defined on the set of all positive integers. We shall be concerned, almost always, with integer-valued arithmetical functions.

The *Möbius function* μ is defined, for any positive integer n, by the following three properties:

(i) $\mu(1)=1$;
(ii) $\mu(n)=(-1)^k$, if n is a product of k different primes;
(iii) $\mu(n)=0$, if n is divisible by a square different from 1.

As a simple consequence of the definition, we have

$$\sum_{d|n}\mu(d)=\begin{cases}1, & \text{if } n=1\,,\\ 0, & \text{if } n>1\,,\end{cases} \tag{2}$$

where the summation is over all the positive divisors d of n.

There are two elementary inversion formulas governing the Möbius function:

(The first Möbius inversion formula). If f is an arithmetical function, and

$$g(n)=\sum_{d|n} f(d),$$

then

$$f(n) = \sum_{d|n} \mu(d) g\left(\frac{n}{d}\right), \tag{3}$$

and conversely.

(The second Möbius inversion formula). If f is a function defined for $x \geqslant 1$, *and*

$$g(x) = \sum_{n \leqslant x} f\left(\frac{x}{n}\right),$$

then

$$f(x) = \sum_{n \leqslant x} \mu(n) g\left(\frac{x}{n}\right), \quad \textit{for} \quad x \geqslant 1, \tag{4}$$

and conversely.

If we set $f(x)=1$, $f(x)=x$, $f(x)=x\log x$, $f(x)=x\log^2 x$, in formula (4), and note that

$$\sum_{n \leqslant x} \frac{1}{n} = \log x + \gamma + O\left(\frac{1}{x}\right), \tag{5}$$

where γ is Euler's constant, while

$$\begin{aligned} \sum_{n \leqslant x} \frac{\log n}{n} &= \frac{1}{2}\log^2 x + c_1 + O\left(\frac{\log x}{x}\right), \\ \sum_{n \leqslant x} \frac{\log^2 n}{n} &= \frac{1}{3}\log^3 x + c_2 + O\left(\frac{\log^2 x}{x}\right), \\ \sum_{n \leqslant x} \log \frac{x}{n} = O(x), &\quad \sum_{n \leqslant x} \log^2 \frac{x}{n} = O(x), \end{aligned} \tag{6}$$

where c_1 and c_2 are constants, we successively obtain the following estimates:

$$\sum_{n \leqslant x} \frac{\mu(n)}{n} = O(1), \quad \sum_{n \leqslant x} \frac{\mu(n)}{n} \log \frac{x}{n} = O(1), \sum_{n \leqslant x} \frac{\mu(n)}{n} \log^2 \frac{x}{n} = 2\log x + O(1), \tag{7}$$

and

$$\sum_{n \leqslant x} \frac{\mu(n)}{n} \log^3 \frac{x}{n} = 3\log^2 x - 6\gamma \log x + O(1). \tag{8}$$

The *von Mangoldt function* Λ is defined, for any positive integer n, by the requirement that

$$\Lambda(n) = \begin{cases} \log p, & \text{if } n \text{ is a prime power } p^m, m>0, \\ 0, & \text{otherwise.} \end{cases} \tag{9}$$

As a simple consequence of the definition, we have

$$\sum_{d|n} \Lambda(d) = \log n, \tag{10}$$

where d runs through all the positive divisors of n. Formula (10), in conjunction with the first Möbius inversion formula, gives

$$\Lambda(n) = \sum_{d|n} \mu(d) \log \frac{n}{d}. \tag{11}$$

In view of (2), this can be written as

$$\Lambda(n) = -\sum_{d|n} \mu(d) \log d. \tag{12}$$

We recall the definition of *Chebyshev's functions* ϑ and ψ. If p denotes a prime, and m a positive integer, and $x>0$, then

$$\vartheta(x) = \sum_{p \leqslant x} \log p, \tag{13}$$

and

$$\psi(x) = \sum_{p^m \leqslant x} \log p. \tag{14}$$

It is immediate that

$$\psi(x) = \sum_{n \leqslant x} \Lambda(n), \quad x \geqslant 1. \tag{15}$$

The following variant of the second Möbius inversion formula, due to K. Iseki and T. Tatuzawa, can be used to prove Selberg's formula: *If F is a function defined for $x \geqslant 1$, and*

$$G(x) = \sum_{n \leqslant x} F\left(\frac{x}{n}\right) \log x,$$

then

$$F(x) \log x + \sum_{n \leqslant x} F\left(\frac{x}{n}\right) \Lambda(n) = \sum_{n \leqslant x} \mu(n) G\left(\frac{x}{n}\right). \tag{16}$$

For, by the definition of G, we have

$$\sum_{n \leqslant x} \mu(n) G\left(\frac{x}{n}\right) = \sum_{n \leqslant x} \mu(n) \sum_{m \leqslant \frac{x}{n}} F\left(\frac{x}{mn}\right) \log \frac{x}{n},$$

and by interchanging the order of summation on the right-hand side, we have

$$\sum_{n\leqslant x}\mu(n)\,G\left(\frac{x}{n}\right)=\sum_{N\leqslant x}F\left(\frac{x}{N}\right)\sum_{d|N}\mu(d)\log\frac{x}{d}$$
$$=\sum_{N\leqslant x}F\left(\frac{x}{N}\right)\sum_{d|N}\mu(d)\left(\log\frac{x}{N}+\log\frac{N}{d}\right)$$
$$=F(x)\log x+\sum_{n\leqslant x}F\left(\frac{x}{n}\right)\Lambda(n),$$

if we use (2) and (11). This proves (16).

We are now in a position to prove

THEOREM 1 (SELBERG'S FORMULA). *If* $x\geqslant 1$, *we have*

$$\psi(x)\log x+\sum_{n\leqslant x}\psi\left(\frac{x}{n}\right)\Lambda(n)=2x\log x+O(x). \tag{17}$$

PROOF. If we take $F(x)=\psi(x)-x+\gamma+1$, where γ is Euler's constant, in formula (16), we get

$$G(x)=\sum_{n\leqslant x}\left(\psi\left(\frac{x}{n}\right)-\frac{x}{n}+\gamma+1\right)\log x. \tag{18}$$

In order to estimate $G(x)$, we observe that, by (15), we have

$$\sum_{n\leqslant x}\psi\left(\frac{x}{n}\right)=\sum_{n\leqslant x}\sum_{m\leqslant\frac{x}{n}}\Lambda(m)=\sum_{mn\leqslant x}\Lambda(m)=\sum_{r\leqslant x}\sum_{d|r}\Lambda(d)=\sum_{r\leqslant x}\log r\quad\text{(by (10))}$$
$$=\int_1^x\log\xi\,d\xi+O(\log x)=x\log x-x+O(\log x). \tag{19}$$

From (5) and (19) we have

$$\sum_{n\leqslant x}\psi\left(\frac{x}{n}\right)\log x=x\log^2 x-x\log x+O(\log^2 x),$$
$$\sum_{n\leqslant x}\frac{x}{n}\log x=x\log^2 x+\gamma x\log x+O(\log x),$$

while trivially

$$\sum_{n\leqslant x}(\gamma+1)\log x=(\gamma+1)x\log x+O(\log x),$$

so that by (18) we obtain the estimate

$$G(x)=O(\log^2 x).$$

If we use this on the right-hand side of (16), we get

$$\left|\sum_{n\leqslant x}\mu(n)G\left(\frac{x}{n}\right)\right|\leqslant\sum_{n\leqslant x}\left|G\left(\frac{x}{n}\right)\right|=O(x),\quad\text{by (6)}.$$

Thus formula (16) leads to

$$F(x)\log x+\sum_{n\leqslant x}F\left(\frac{x}{n}\right)\Lambda(n)=O(x),\tag{20}$$

with $F(x)=\psi(x)-x+\gamma+1$. Now

$$\sum_{n\leqslant x}\Lambda(n)\left[\frac{x}{n}\right]=\sum_{n\leqslant x}\Lambda(n)\sum_{m\leqslant\frac{x}{n}}1=\sum_{mn\leqslant x}\Lambda(n),$$

where $\left[\frac{x}{n}\right]$ denotes the greatest integer less than or equal to $\frac{x}{n}$, so that, by (19), we have

$$\sum_{n\leqslant x}\Lambda(n)\left[\frac{x}{n}\right]=x\log x+O(x).$$

This, in turn, leads to the formula

$$\sum_{n\leqslant x}\Lambda(n)\left(\frac{x}{n}\right)=\sum_{n\leqslant x}\Lambda(n)\left[\frac{x}{n}\right]+O\left(\sum_{n\leqslant x}\Lambda(n)\right)=x\log x+O(x)+O(\psi(x)).\tag{21}$$

If we use (21) in (20), we get

$$\begin{aligned}&\psi(x)\log x+\sum_{n\leqslant x}\psi\left(\frac{x}{n}\right)\Lambda(n)\\&=x\log x+x\sum_{n\leqslant x}\frac{\Lambda(n)}{n}-(\gamma+1)\log x-(\gamma+1)\sum_{n\leqslant x}\Lambda(n)+O(x)\\&=2x\log x+O(x)+O(\psi(x)).\end{aligned}\tag{22}$$

Since ψ and Λ are non-negative functions, this implies that

$$\psi(x)(\log x+O(1))\leqslant O(x\log x),$$

which, in turn, implies that

$$\psi(x)=O(x),\tag{23}$$

and, if we use this in (22), we get the theorem.

REMARKS. (i) By combining (23) and (21), we get the formula

$$\sum_{n\leqslant x}\frac{\Lambda(n)}{n}=\log x+O(1),\tag{24}$$

and by combining (23) and (24), we have

$$\int_1^x \frac{\psi(t)}{t^2}\,dt = \log x + O(1), \qquad x \geqslant 1. \tag{25}$$

For

$$\int_1^x \frac{\psi(t)}{t^2}\,dt = \int_1^x \sum_{n\leqslant t} \Lambda(n)\cdot\frac{dt}{t^2} = \sum_{n\leqslant x}\Lambda(n)\int_n^x \frac{dt}{t^2} = \sum_{n\leqslant x}\Lambda(n)\left(\frac{1}{n}-\frac{1}{x}\right)$$

$$= \sum_{n\leqslant x}\frac{\Lambda(n)}{n} - \frac{\psi(x)}{x}.$$

(ii) Selberg's formula leads to an important inequality for the function defined by

$$\bar{r}(x) = e^{-x}\psi(e^x) - 1, \qquad x > 0. \tag{26}$$

It can be proved that, for $x > 1$, we have

$$|\bar{r}(x)| \leqslant \frac{1}{x}\int_0^x |\bar{r}(t)|\,dt + O\left(\frac{\log x}{x}\right). \tag{27}$$

(iii) It is well known that the prime number theorem as stated in (1) is equivalent to either of the assertions that

$$\lim_{x\to\infty}\frac{\psi(x)}{x} = 1, \quad \text{or} \quad \lim_{x\to\infty}\frac{\vartheta(x)}{x} = 1, \tag{28}$$

the equivalence being provable by elementary arguments dating from Chebyshev. Clearly (28) is equivalent to the assertion that $\lim_{x\to\infty}\bar{r}(x) = 0$. It can be shown that (27) implies this and hence the prime number theorem. We shall however follow a different path.

§ 2. A variant of Selberg's formula. Let

$$r_n = \begin{cases} \dfrac{1}{n} - \dfrac{\Lambda(n)}{n}, & \text{for } n = 2, 3, \ldots, \\ 1 - 2\gamma, & \text{for } n = 1, \end{cases} \tag{29}$$

and

$$r(x) = \begin{cases} \sum_{n\leqslant x} r_n, & \text{for } x \geqslant 1, \\ 0, & \text{for } x < 1. \end{cases} \tag{30}$$

A variant of formula (17) is given by

THEOREM 2 (E. WIRSING). *If* $x \geqslant 1$, *then*

$$r(x)\log x - \sum_{k \leqslant x} r_k r\left(\frac{x}{k}\right) = O(1), \tag{31}$$

and

$$\left|\sum_{y < n \leqslant x} r_n\right| \leqslant \log\frac{x}{y} + O\left(\frac{1}{\log y}\right), \qquad y > 1. \tag{32}$$

PROOF. If we set $F(x) = x r(x)$ in formula (16), we obtain

$$r(x)\log x + \sum_{n \leqslant x} \frac{\Lambda(n)}{n} r\left(\frac{x}{n}\right) = \sum_{mn \leqslant x} r\left(\frac{x}{mn}\right) \frac{\mu(n)}{mn} \log\frac{x}{n}. \tag{33}$$

In order to prove that the sum on the right-hand side is $O(1)$, we need an estimate of $s(x)$ defined by

$$s(x) = \sum_{m \leqslant x} \frac{1}{m} r\left(\frac{x}{m}\right), \qquad x \geqslant 1. \tag{34}$$

It is obvious from (10) that

$$s(x) = \sum_{mn \leqslant x} \frac{1}{m} r_n = \sum_{k \leqslant x} \frac{1}{k} \sum_{n|k} n r_n = \sum_{k \leqslant x} \frac{1}{k}(d(k) - \log k - 2\gamma), \tag{35}$$

where $d(k)$ denotes the number of positive divisors of the positive integer k.

From Dirichlet's elementary estimate

$$\sum_{n \leqslant x} d(n) = x\log x + (2\gamma - 1)x + O(\sqrt{x}),$$

we have

$$\sum_{n \leqslant x} (d(n) - \log n - 2\gamma) = O(x^{\frac{1}{2}}), \tag{36}$$

whence, by partial summation in (35), we get the estimate

$$s(x) = c + O\left(\frac{1}{\sqrt{x}}\right), \qquad x \geqslant 1, \tag{37}$$

where c is a constant. Further we have

$$\sum_{n \leqslant x} \frac{1}{\sqrt{nx}} \log\frac{x}{n} = O(1), \tag{38}$$

if we write $\frac{2}{x} \sum_{n \leqslant x} \sqrt{\frac{x}{n}} \log\sqrt{\frac{x}{n}}$ as a Stieltjes integral.

Now the right-hand side of (33) is

$$\sum_{mn\leqslant x} r\left(\frac{x}{mn}\right)\frac{\mu(n)}{mn}\log\frac{x}{n}=\sum_{n\leqslant x}\frac{\mu(n)}{n}\cdot\log\frac{x}{n}\cdot s\left(\frac{x}{n}\right)=O(1),$$

by (34), (37), (38), and the second estimate in (7). Hence (33) gives

$$r(x)\log x+\sum_{n\leqslant x}\frac{\Lambda(n)}{n}r\left(\frac{x}{n}\right)=O(1). \tag{39}$$

Noting that

$$\sum_{k\leqslant x}\frac{1}{k}r\left(\frac{x}{k}\right)=s(x)=O(1),$$

by (37), and that

$$2\gamma r(x)=O(1),$$

since

$$r(x)=O(1), \tag{40}$$

by (5) and (24), we deduce (31) from (39) and (29).

To prove (32), we note that if $y\geqslant 1$, and $x>y$, then

$$r(x)-r(y)\leqslant\sum_{y<n\leqslant x}\frac{1}{n}=\log\frac{x}{y}+O\left(\frac{1}{y}\right). \tag{41}$$

If $y<x<Ky$, we have

$$(r(x)-r(y))\log x=r(x)\log x-r(y)\log y-r(y)\log\frac{x}{y},$$

and if we use (39) and (40), and note that $r\left(\frac{y}{k}\right)=0$ for $k>y$, we obtain

$$(r(x)-r(y))\log x=-\sum_{k\leqslant x}\frac{\Lambda(k)}{k}\left(r\left(\frac{x}{k}\right)-r\left(\frac{y}{k}\right)\right)+O(1),$$

where the $O(1)$ depends on K. If we now apply (41), we get

$$\begin{aligned}(r(x)-r(y))\log x&\geqslant-\sum_{k\leqslant x}\frac{\Lambda(k)}{k}\left\{\log\frac{x}{y}+O\left(\frac{k}{y}\right)\right\}+O(1)\\&=-\log x\cdot\log\frac{x}{y}+O(1),\end{aligned}$$

by (24) and (23). Hence

$$r(x)-r(y)\geqslant -\log\frac{x}{y}+O\left(\frac{1}{\log x}\right), \tag{42}$$

if $1\leqslant y<x<Ky$. If K is sufficiently large, and $x\geqslant Ky$, then (42) follows directly from (40).

We deduce (32) from (41) and (42), and the proof of Theorem 2 is complete.

Using Stieltjes integrals instead of sums, we can rewrite (31) as

$$r(x)\log x=\int_{1-}^{x} r\left(\frac{x}{t}\right)dr(t)+O(1), \quad x\geqslant 1, \tag{43}$$

noting that $r(t)$ is continuous on the right, but discontinuous on the left, at $t=1$. If $r(t)$ and $r\left(\frac{x}{t}\right)$ are discontinuous for the same t, the integral in (43) is not defined. Since that is possible, however, only for integral values of x, we exclude such values of x in what follows. This does not affect the analysis.

If we set

$$\xi=\log x, \qquad \sum_{n\leqslant e^{\xi}} r_n=\rho(\xi), \tag{44}$$

then (43) and (32) lead to

THEOREM 2′. *If* $\xi\geqslant 0$, $\xi\neq\log n$, $n=1,2,\ldots$, *then*

$$\xi\rho(\xi)=\int_{0-}^{\xi}\rho(\xi-\eta)\,d\rho(\eta)+O(1). \tag{45}$$

There exists a constant K, such that

$$|\rho(\xi)-\rho(\eta)|\leqslant\xi-\eta+\frac{K}{1+\eta}, \quad \textit{for} \quad 0\leqslant\eta\leqslant\xi. \tag{46}$$

The function ρ can be approximated by a continuous, piece-wise linear function σ, as shown by the following

THEOREM 3 (E. WIRSING). *Given* $\varepsilon>0$, *there exists a continuous, piece-wise linear function* σ, *such that*

$$\xi\sigma(\xi)=\int_{0}^{\xi}\sigma(\xi-\eta)\,d\sigma(\eta)+O(\log\xi), \quad \xi\geqslant 2, \tag{47}$$

and if σ' denotes the derivative of σ, then

$$|\sigma'(\xi)| \leqslant 1+\varepsilon, \tag{48}$$

for $\xi>0$, except for a countable set of points $\xi=\xi_n$, $n=1,2,\ldots$. Further

$$|\sigma(\xi)-\rho(\xi)| = O\left(\frac{1}{1+\xi}\right), \qquad \xi \geqslant 0, \tag{49}$$

where ρ is defined as in (44).

Proof. Given $\varepsilon>0$, ε being sufficiently small, we define the sequence (ξ_n), $n=0,1,2,\ldots$, as follows:

$$\xi_0=0, \qquad \xi_{\nu+1}=\xi_\nu+\frac{K}{\varepsilon(1+\xi_\nu)}. \tag{50}$$

It is easy to see that

$$0<\xi_1<\xi_2<\cdots\to\infty.$$

We define the function σ by the requirement that

$$\begin{aligned} \sigma(\xi)=0 \quad \text{for} \quad \xi<0; \qquad \sigma(0)=0=\rho(0-); \\ \sigma(\xi_\nu)=\rho(\xi_\nu), \qquad \nu=1,2,\ldots; \end{aligned} \tag{51}$$

and σ is linear in $(\xi_\nu, \xi_{\nu+1})$, $\nu=0,1,2,\ldots$. Then for $\xi_\nu<\xi<\xi_{\nu+1}$, $\nu\geqslant 1$, we have

$$\begin{aligned} |\sigma'(\xi)| &= \left|\frac{\sigma(\xi_{\nu+1})-\sigma(\xi_\nu)}{\xi_{\nu+1}-\xi_\nu}\right| = \left|\frac{\rho(\xi_{\nu+1})-\rho(\xi_\nu)}{\xi_{\nu+1}-\xi_\nu}\right| \\ &\leqslant 1+\frac{K}{(1+\xi_\nu)(\xi_{\nu+1}-\xi_\nu)} = 1+\varepsilon. \qquad \text{(by (46))} \end{aligned}$$

If $0=\xi_0<\xi<\xi_1$, and C is a constant such that $|\rho(\xi)|<C$, (see (40)), then

$$|\sigma'(\xi)| = \left|\frac{\sigma(\xi_1)}{\xi_1}\right| = \left|\frac{\rho(\xi_1)}{\xi_1}\right| = \left|\rho\left(\frac{K}{\varepsilon}\right)\cdot\frac{\varepsilon}{K}\right| < \frac{C}{K}\cdot\varepsilon<1+\varepsilon,$$

if ε is sufficiently small. Hence

$$|\sigma'(\xi)|\leqslant 1+\varepsilon, \qquad \xi>0,\ \xi\neq\xi_1,\xi_2,\ldots,$$

which proves (48).

To prove (49), we observe that for $\xi_\nu < \xi < \xi_{\nu+1}$, $\nu \geqslant 0$, we have

$$\begin{aligned}|\sigma(\xi) - \rho(\xi)| &= |\sigma(\xi) - \sigma(\xi_\nu) + \rho(\xi_\nu) - \rho(\xi)| \\ &\leqslant (\xi - \xi_\nu)(1+\varepsilon) + \xi - \xi_\nu + \frac{K}{1+\xi_\nu} \\ &\leqslant (\xi_{\nu+1} - \xi_\nu)(2+\varepsilon) + \frac{K}{1+\xi_\nu},\end{aligned}$$

which implies (49). We shall use this to replace ρ by σ in (45).

Since

$$\sum_{y<n\leqslant x} |r_n| \leqslant \sum_{y<n\leqslant x} \frac{1}{n}(1+\Lambda(n)) = 2\log\frac{x}{y} + O(1), \quad y \geqslant 1,$$

by (5) and (24), we have

$$R(\eta) = O(1) + O(\xi - \eta), \qquad 0 \leqslant \eta \leqslant \xi, \tag{52}$$

where

$$R(\eta) = \int_\eta^\xi |d\rho(t)|.$$

Now, if $\xi \geqslant 2$, then

$$\begin{aligned}\int_{0-}^{\xi} (\rho(\xi-\eta) - \sigma(\xi-\eta))\, d\rho(\eta) &= O\left(\int_{0-}^{\xi} \frac{1}{1+\xi-\eta} |d\rho(\eta)|\right) && \text{(by (49))} \\ &= O\left(\frac{R(0-)}{\xi}\right) + O\left(\int_0^\xi \frac{R(\eta)\, d\eta}{(1+\xi-\eta)^2}\right) \\ &= O(1) + O\left(\int_0^\xi \frac{d\eta}{1+\xi-\eta}\right) && \text{(by (52))} \\ &= O(\log \xi). && (53)\end{aligned}$$

Further

$$\begin{aligned}\int_{0-}^{\xi} \sigma(\xi-\eta)\, d(\rho(\eta) - \sigma(\eta)) &= \int_0^\xi (\rho(\eta) - \sigma(\eta))\sigma'(\xi-\eta)\, d\eta \\ &= O\left(\int_0^\xi \frac{d\eta}{1+\eta}\right) \\ &= O(\log \xi). && (54)\end{aligned}$$

By (53) and (54), we have

$$\int_{0-}^{\xi} \rho(\xi-\eta)\,d\rho(\eta) = \int_{0}^{\xi} \sigma(\xi-\eta)\,d\sigma(\eta)+O(\log\xi)\,. \tag{55}$$

But by (49) we have $\xi\rho(\xi)=\xi\sigma(\xi)+O(1)$. Hence (45) leads to the formula (47). This completes the proof of the theorem.

§ 3. Wirsing's inequality. The prime number theorem follows from Theorem 3 when combined with the following

THEOREM 4 (E. WIRSING). *Let f and g be real-valued Lebesgue measurable functions defined on $[0,\infty)$, which are square-integrable over every finite interval. Let*

$$\overline{\lim_{x\to\infty}}\, \frac{1}{x}\int_0^x f^2(y)\,dy=F, \qquad \overline{\lim_{x\to\infty}}\, \frac{1}{x}\int_0^x g^2(y)\,dy=G, \tag{56}$$

where F and G are finite. Let the function h, defined by the relation

$$h(x) = \frac{1}{x}\int_0^x f(x-y)g(y)\,dy\,, \quad x>0\,, \tag{57}$$

satisfy the condition

$$\lim_{x\to\infty} \frac{1}{x}\int_0^x h(y)\,dy=0\,. \tag{58}$$

Then

$$\overline{\lim_{x\to\infty}}\, \frac{1}{x}\int_{x_0}^x h^2(y)\,dy\leqslant\tfrac{1}{2}FG\,, \tag{59}$$

where x_0 is any number greater than zero.

PROOF. If $FG=0$, the conclusion follows from Schwarz's inequality. We may therefore assume, without loss of generality, that $F=G=1$.

Instead of assuming (56), let us first assume that

$$\int_0^x f^2(y)\,dy\leqslant x\,, \qquad \int_0^x g^2(y)\,dy\leqslant x\,. \tag{60}$$

Define

$$P(x)=P(f,g;x) = \int_0^x f(x-y)g(y)\,dy\,. \tag{61}$$

Then

$$P^2(x) \leqslant \int_0^x f^2(x-y)\,dy \cdot \int_0^x g^2(y)\,dy \leqslant x^2,$$

so that

$$|h(x)| = \left|\frac{P(x)}{x}\right| \leqslant 1, \quad \text{for} \quad x>0. \tag{62}$$

Since f and g are square-integrable in every finite interval, the integral defining $P(x)$ exists, and the function $P(x)$ is continuous, hence also $h(x)$.

Let N be a sufficiently large positive integer, and

$$\varepsilon = \frac{1}{N}, \qquad x_n = (1+\varepsilon)^n, \qquad n = 1, 2, \ldots. \tag{63}$$

We seek to estimate the integral $\int_{x_n}^{x_{n+1}} h^2(x)\,dx$ and use it to prove (59). For that purpose we define

$$k(a) = \frac{1}{x_{n+1}} \int_0^{x_{n+1}-a} f(y+a) f(y)\,dy, \quad 0 \leqslant a \leqslant x_{n+1}. \tag{64}$$

Then k is a continuous function of a.

Now, if $x_n \leqslant x \leqslant x' \leqslant x_{n+1}$, $x' = x+a$, and $\lambda > 0$, then

$$\lambda P(x') \pm \frac{1}{\lambda} P(x) = \int_0^x \left(\lambda f(x'-y) \pm \frac{1}{\lambda} f(x-y)\right) g(y)\,dy$$
$$+ \lambda \int_x^{x'} f(x'-y) g(y)\,dy.$$

If we set

$$\varphi(y) = \begin{cases} \lambda f(x'-y) \pm \dfrac{1}{\lambda} f(x-y), & 0 \leqslant y \leqslant x, \\ \lambda f(x'-y), & x < y \leqslant x', \end{cases}$$

then

$$\lambda P(x') \pm \frac{1}{\lambda} P(x) = \int_0^{x'} \varphi(y) g(y)\,dy,$$

so that

$$\left(\lambda P(x') \pm \frac{1}{\lambda} P(x)\right)^2 \leqslant \int_0^{x'} g^2(y)\,dy \left(\int_0^{x} \left(\lambda f(x'-y) \pm \frac{1}{\lambda} f(x-y)\right)^2 dy + \lambda^2 \int_x^{x'} f^2(x'-y)\,dy \right)$$

$$\leqslant x' \left(\int_0^{x_{n+1}-a} \left(\lambda f(t+a) \pm \frac{1}{\lambda} f(t)\right)^2 dt + \lambda^2 \int_0^a f^2(t)\,dt \right)$$

$$\leqslant x_{n+1} \left(\left(\lambda^2 + \frac{1}{\lambda^2}\right) \int_0^{x_{n+1}} f^2(t)\,dt \pm 2 \int_0^{x_{n+1}-a} f(t+a) f(t)\,dt + \lambda^2 a \right) \quad \text{(by (60))}$$

$$\leqslant x_{n+1}^2 \left(\lambda^2 + \frac{1}{\lambda^2} \pm 2k(a)\right) + x_{n+1} \lambda^2 a.$$

If we substitute $x h(x)$ for $P(x)$, and note that $x'-x=a$, $x_n \leqslant x \leqslant x' \leqslant x_{n+1}$, $x_{n+1}-x_n=\varepsilon x_n$, $|h(x)| \leqslant 1$, $|k(a)| \leqslant 1$, we get, after some simplification,

$$\left(\lambda h(x') \pm \frac{1}{\lambda} h(x)\right)^2 \leqslant \lambda^2 + \frac{1}{\lambda^2} \pm 2k(a) + K\varepsilon\left(\lambda^2 + \frac{1}{\lambda^2}\right),$$

where K is a positive, absolute constant. This can be rewritten as

$$\pm 2(h(x')h(x) - k(a)) \leqslant \lambda^2(1 + K\varepsilon - h^2(x')) + \frac{1}{\lambda^2}(1 + K\varepsilon - h^2(x)).$$

If we now choose

$$\lambda^2 = (1 + K\varepsilon - h^2(x))^{\frac{1}{2}} (1 + K\varepsilon - h^2(x'))^{-\frac{1}{2}},$$

we obtain the inequality

$$|h(x)h(x') - k(a)| \leqslant (1 + K\varepsilon - h^2(x))^{\frac{1}{2}} (1 + K\varepsilon - h^2(x'))^{\frac{1}{2}}. \tag{65}$$

We shall use this to estimate $\int_{x_n}^{x_{n+1}} h^2(x)\,dx$. We distinguish two cases.

CASE (i). There exist points $x, x' \in [x_n, x_{n+1}]$, such that $x' = x + a > x$, $|x' - x| \leqslant \frac{1}{N}(x_{n+1} - x_n)$, $h(x)h(x') \leqslant 0$, for which $h^2(x) + h^2(x') \geqslant 1 + K\varepsilon$.

Then, by (65), we have

$$k(a) - h(x)h(x') \leqslant [(1+K\varepsilon)^2 - (1+K\varepsilon)(h^2(x)+h^2(x')) + h^2(x)h^2(x')]^{\frac{1}{2}}$$
$$\leqslant |h(x)\cdot h(x')|,$$

so that $k(a) \leqslant 0$. By definition (64), however, $k(0) \geqslant 0$; and k is continuous. Hence there exists a number a_0, $0 \leqslant a_0 \leqslant a$, such that $k(a_0) = 0$. If $y \in [x_n, x_{n+1} - a_0]$, then by (65),

$$|h(y)h(y+a_0)| \leqslant (1+K\varepsilon - h^2(y))^{\frac{1}{2}}(1+K\varepsilon - h^2(y+a_0))^{\frac{1}{2}},$$

or

$$h^2(y)h^2(y+a_0) \leqslant (1+K\varepsilon)^2 - (1+K\varepsilon)(h^2(y)+h^2(y+a_0)) + h^2(y)h^2(y+a_0),$$

or

$$h^2(y) + h^2(y+a_0) \leqslant 1 + K\varepsilon.$$

If we use this inequality, together with the fact that $h^2(x) \leqslant 1$, we have

$$2\int_{x_n}^{x_{n+1}} h^2(y)\,dy = \int_{x_n}^{x_{n+1}-a_0} (h^2(y)+h^2(y+a_0))\,dy + \int_{x_n}^{x_n+a_0} h^2(y)\,dy + \int_{x_{n+1}-a_0}^{x_{n+1}} h^2(y)\,dy$$
$$\leqslant (1+K\varepsilon)(x_{n+1}-x_n) + 2a_0$$
$$\leqslant \left(1 + K\varepsilon + \frac{2}{N}\right)(x_{n+1}-x_n),$$

since $a \leqslant \frac{1}{N}(x_{n+1}-x_n)$. It follows that

$$\int_{x_n}^{x_{n+1}} h^2(y)\,dy \leqslant (\tfrac{1}{2} + K_1\varepsilon)(x_{n+1}-x_n), \qquad K_1 = \tfrac{1}{2}K + 1, \tag{66}$$

since $\varepsilon = \frac{1}{N}$.

CASE (ii). For every two points $x, x' \in [x_n, x_{n+1}]$, such that $|x' - x| \leqslant \frac{1}{N}(x_{n+1} - x_n)$, $h(x)h(x') \leqslant 0$, we have $h^2(x) + h^2(x') < 1 + K\varepsilon$.

We define

$$x_{n,\nu} = \frac{\nu}{N}(x_{n+1} - x_n) + x_n, \qquad \nu = 0, 1, 2, \ldots, N,$$

and consider the interval $[x_{n,\nu-1}, x_{n,\nu}]$, $\nu \geqslant 1$. Since

$$x_{n,\nu} - x_{n,\nu-1} = \frac{1}{N}(x_{n+1} - x_n) = \frac{\varepsilon}{N} x_n,$$

it follows from assumption (58) that

$$\left| \int_{x_{n,\nu-1}}^{x_{n,\nu}} h(y)\,dy \right| \leqslant \varepsilon(x_{n,\nu} - x_{n,\nu-1}), \quad \text{for} \quad n \geqslant n_0(\varepsilon). \tag{67}$$

We consider separately the case when h does not change sign in the interval $[x_{n,\nu-1}, x_{n,\nu}]$ from the case when it does.

If h does not change sign in $[x_{n,\nu-1}, x_{n,\nu}]$, then, since $|h(y)| \leqslant 1$, we have

$$\int_{x_{n,\nu-1}}^{x_{n,\nu}} h^2(y)\,dy \leqslant \int_{x_{n,\nu-1}}^{x_{n,\nu}} |h(y)|\,dy = \left| \int_{x_{n,\nu-1}}^{x_{n,\nu}} h(y)\,dy \right| \leqslant \varepsilon(x_{n,\nu} - x_{n,\nu-1}) \quad \text{(by (67))}$$

$$\leqslant \tfrac{1}{2}(x_{n,\nu} - x_{n,\nu-1}). \tag{68}$$

If, on the other hand, h changes sign in $[x_{n,\nu-1}, x_{n,\nu}]$, then let

$$s_1 = \sup h(y), \qquad s_2 = \inf h(y), \qquad x_{n,\nu-1} \leqslant y \leqslant x_{n,\nu}.$$

Obviously $-1 \leqslant s_2 < 0 < s_1 \leqslant 1$, and by our assumptions in Case (ii), $s_1^2 + s_2^2 < 1 + K\varepsilon$. Hence

$$-s_1 s_2 \leqslant \tfrac{1}{2}(s_1^2 + s_2^2) \leqslant \tfrac{1}{2} + \tfrac{1}{2}K\varepsilon,$$

while

$$|h(y) - \tfrac{1}{2}(s_1 + s_2)| \leqslant \tfrac{1}{2}(s_1 - s_2),$$

which implies that

$$h^2(y) - (s_1 + s_2)h(y) + (\tfrac{1}{2}(s_1 + s_2))^2 \leqslant (\tfrac{1}{2}(s_1 - s_2))^2,$$

or that

$$h^2(y) \leqslant -s_1 s_2 + (s_1 + s_2)h(y)$$
$$\leqslant \tfrac{1}{2} + \tfrac{1}{2}K\varepsilon + (s_1 + s_2)h(y).$$

By (67) it follows that

$$\int_{x_{n,\nu-1}}^{x_{n,\nu}} h^2(y)\,dy \leqslant (\tfrac{1}{2} + \tfrac{1}{2}K\varepsilon + |s_1 + s_2|\varepsilon)\,(x_{n,\nu} - x_{n,\nu-1})$$
$$\leqslant (\tfrac{1}{2} + K_2\varepsilon)\,(x_{n,\nu} - x_{n,\nu-1}), \qquad K_2 = K_1 + 1. \tag{69}$$

If in (69) and (68) we sum over ν, from $\nu = 1$ to $\nu = N$, we obtain (66) also in Case (ii), with K_2 in place of K_1.

Thus we have altogether

$$\int_{x_n}^{x_{n+1}} h^2(y)\,dy \leqslant (\tfrac{1}{2}+K_2\varepsilon)(x_{n+1}-x_n), \qquad n \geqslant n_0(\varepsilon), \tag{70}$$

where K_2 is a positive, absolute constant.

For arbitrary $x \geqslant x_{n_0(\varepsilon)}$, we choose n such that $x_n \leqslant x < x_{n+1}$. Then, because of (70), and of the fact that $h^2(x) \leqslant 1$, we have for *any* fixed $x_0 > 0$,

$$\int_{x_0}^{x} h^2(y)\,dy \leqslant x_{n_0(\varepsilon)} + (\tfrac{1}{2}+K_2\varepsilon)(x_n - x_{n_0(\varepsilon)}) + x - x_n \leqslant x_{n_0(\varepsilon)} + (\tfrac{1}{2}+K_3\varepsilon)x,$$

where $K_3 = K_2 + 1$. This implies (59), and hence the theorem is proved under hypothesis (60) instead of (56), with $F = G = 1$.

We now revert to condition (56) with $F = G = 1$. Given $\varepsilon > 0$, we can then choose $x_1(\varepsilon)$, such that

$$\int_0^x f^2(y)\,dy \leqslant (1+\varepsilon)x, \qquad \int_0^x g^2(y)\,dy \leqslant (1+\varepsilon)x,$$

for $x \geqslant x_1$. Let $f = f_1 + f_2$, where $f_1(x) = 0$ for $x < x_1$, and $f_2(x) = 0$ for $x \geqslant x_1$; and similarly let $g = g_1 + g_2$. We can then apply the result we have already obtained to the functions f_1, g_1, and $h_1(x) = \dfrac{1}{x} P(f_1, g_1; x)$, and infer that

$$\int_{x_0}^{x} h_1^2(y)\,dy \leqslant \tfrac{1}{2}(1+\varepsilon)^2 x + o(x).$$

But by Schwarz's inequality,

$$P(f, g; x) = P(f_1, g_1; x) + O(\sqrt{x}).$$

Hence

$$\int_{x_0}^{x} h^2(y)\,dy = \int_{x_0}^{x} h_1^2(y)\,dy + O(\sqrt{x}) \leqslant \tfrac{1}{2}(1+\varepsilon)^2 x + o(x),$$

which proves (59), and hence the theorem.

§ 4. The prime number theorem. Combining Theorems 3 and 4 we can deduce

THEOREM 5.
$$\lim_{x\to\infty} \frac{\psi(x)}{x} = 1,$$

where ψ is Chebyshev's function.

PROOF. By Theorem 3, we have, for a sufficiently small $\varepsilon>0$, and $x\geqslant 2$,

$$\sigma(x)=\frac{1}{x}\int_0^x \sigma(x-y)\sigma'(y)dy+O\left(\frac{1}{\sqrt{x}}\right), \tag{71}$$

where σ is differentiable almost everywhere, and

$$|\sigma'(x)|\leqslant 1+\varepsilon. \tag{72}$$

By (40), $r(x)$ is bounded, so is $\rho(x)$, and $\sigma(x)$. We shall see that

$$\int_0^x \rho(\xi)d\xi=\int_1^{\exp x}\frac{r(t)}{t}dt=O(1). \tag{73}$$

Clearly

$$\int_1^x \frac{r(t)}{t}dt=\sum_{n\leqslant x} r_n\log\frac{x}{n}=\sum_{n\leqslant x}\left(\frac{1}{n}-\frac{\Lambda(n)}{n}\right)\log\frac{x}{n}-2\gamma\log x, \qquad x\geqslant 1. \tag{74}$$

By (5) and (6) we have

$$\sum_{n\leqslant x}\frac{1}{n}\log\frac{x}{n}=\frac{1}{2}\log^2 x+\gamma\log x-c_1+O\left(\frac{\log x}{x}\right). \tag{75}$$

If we put $F(x)=x\log x$ in formula (16), and use (75) as well as (6), (7) and (8), we see that

$$\sum_{n\leqslant x}\frac{\Lambda(n)}{n}\log\frac{x}{n}=\frac{1}{2}\log^2 x-\gamma\log x+O(1). \tag{76}$$

If we use (75) and (76) in (74), we obtain (73). Hence, by (49),

$$\frac{1}{x}\int_0^x \sigma(y)dy=O\left(\frac{\log x}{x}\right)=o(1), \quad \text{as} \quad x\to\infty. \tag{77}$$

We can now apply Wirsing's inequality (Theorem 4) with $f(y)=\sigma(y)$, $g(y)=\sigma'(y)$, and $h(x)=\sigma(x)+o(1)$, and, using (71) and (72), obtain

$$\varlimsup_{x\to\infty}\frac{1}{x}\int_0^x \sigma^2(y)dy\leqslant\frac{1}{2}\varlimsup_{x\to\infty}\frac{1}{x}\int_0^x \sigma^2(y)dy\cdot\varlimsup_{x\to\infty}\frac{1}{x}\int_0^x (\sigma'(y))^2dy$$

$$\leqslant\frac{1}{2}\varlimsup_{x\to\infty}\frac{1}{x}\int_0^x \sigma^2(y)dy\cdot(1+\varepsilon)^2,$$

which implies, if ε is sufficiently small, that

$$\overline{\lim_{x\to\infty}}\frac{1}{x}\int_0^x \sigma^2(y)\,dy=0. \tag{78}$$

Now (71) gives, by Schwarz's inequality,

$$x|\sigma(x)|\leqslant\left(\int_0^x \sigma^2(y)\,dy\right)^{\frac{1}{2}}\left(\int_0^x |\sigma'(y)|^2\,dy\right)^{\frac{1}{2}}+o(x)=o(x),\quad \text{by (78)}.$$

Hence $\sigma(x)=o(1)$, as $x\to\infty$. This implies, by (49), that $\rho(x)=o(1)$, as $x\to\infty$. In other words,

$$\sum_{n\leqslant x}\left(\frac{\Lambda(n)}{n}-\frac{1}{n}\right)+2\gamma=o(1),\quad \text{as}\quad x\to\infty. \tag{79}$$

For $x\geqslant 1$, we have, however,

$$\psi(x)-x=\sum_{n\leqslant x}(\Lambda(n)-1)+O(1)=\sum_{n\leqslant x}\left(\frac{\Lambda(n)-1}{n}\cdot n\right)+O(1).$$

By partial summation, and (79), we obtain

$$\psi(x)-x=-2\gamma x+o(x)+\int_1^x (2\gamma+o(1))\,dt+O(1)=o(x),$$

which proves Theorem 5.

§ 5. The order of magnitude of the divisor function. The arithmetical function $d(n)$, defined as the number of positive divisors of the positive integer n, is called the *divisor function*. If p is a prime, then $d(p)=2$. Hence $\liminf_{n\to\infty} d(n)=2$. By elementary reasoning, one can show that $d(n)=O(n^{\varepsilon})$, for every $\varepsilon>0$. One can also prove that, for every $\delta>0$, there exists a sequence of positive integers (n_k) for which

$$\frac{d(n_k)}{(\log n_k)^{\delta}}\to\infty,\quad \text{as}\quad k\to\infty.$$

With the aid of the prime number theorem, we prove

THEOREM 6. $$\overline{\lim}\,\log d(n)\cdot\frac{\log\log n}{\log n}=\log 2. \tag{80}$$

PROOF. If $n>1$, with the *standard form* $n=\prod_{p|n}p^{a_p}$, then $d(n)=\prod_p(a_p+1)$.

Let the integer m be defined as the product of the first n primes, namely $m=p_1p_2\cdots p_n$. Then $d(m)=2^n$, so that $\log d(m)=n\log 2$. Let $y=p_n$. Then $\log m=\vartheta(y)$ because of (13), and $\vartheta(y)\sim y$ by the prime number theorem, while $n=\pi(p_n)=\pi(y)\sim\dfrac{y}{\log y}$, by the prime number theorem. Hence

$$\log d(m)\sim\log 2\cdot\frac{y}{\log y}\sim\log 2\cdot\frac{\log m}{\log\log m}.$$

Thus, given $\delta>0$, we have, for infinitely many integers m of the type chosen,

$$d(m)>2^{(1-\delta)\log m/(\log\log m)}. \tag{81}$$

To prove an inequality in the opposite direction: given $\delta>0$, we choose ε, such that $0<\varepsilon<\delta$, and η, such that $0<\eta<\dfrac{\varepsilon}{1+\varepsilon}$. Let n be an integer, $n\geqslant 3$. Define

$$\omega=\omega(n)=(1+\varepsilon)\frac{\log 2}{\log\log n},$$

and

$$\Omega=\Omega(n)=(\log n)^{1-\eta}.$$

We then have

$$\Omega^{\omega}=e^{\omega\log\Omega}=e^{(1-\eta)(1+\varepsilon)\log 2}>2, \tag{82}$$

since $(1-\eta)(1+\varepsilon)=1+\varepsilon-\eta(1+\varepsilon)>1$. If n has the standard form $n=\prod_p p^{a_p}$, then

$$\frac{d(n)}{n^{\omega}}=\prod_{p\mid n}\frac{(a_p+1)}{p^{a_p\omega}}=\prod_{\substack{p\leqslant\Omega\\ p\mid n}}\frac{(a_p+1)}{p^{a_p\omega}}\cdot\prod_{\substack{p>\Omega\\ p\mid n}}\frac{(a_p+1)}{p^{a_p\omega}}=\prod\nolimits_1\cdot\prod\nolimits_2,$$

say. Each factor in the product $\prod_2$ is less than or equal to one, for

$$\frac{(a_p+1)}{p^{a_p\omega}}<\frac{(a_p+1)}{\Omega^{a_p\omega}}<\frac{(a_p+1)}{2^{a_p}}\leqslant 1,$$

since $\Omega^{\omega}>2$ by (82), and $(\nu+1)2^{-\nu}\leqslant 1$, for all integers $\nu\geqslant 1$. Thus $\prod_2\leqslant 1$.

If we consider the product $\prod_1$, it has at most $\pi(\Omega)$ factors, where

$$\pi(\Omega)\sim\frac{(\log n)^{1-\eta}}{(1-\eta)\log\log n},$$

by the prime number theorem. Each factor in $\prod_1$ is less than or equal to $((a_p+1)/2^{a_p\omega})<(a_p/2^{a_p\omega})+1$. Further, for fixed ω, the real function

$(u/2^{u\omega})$ has the maximum $1/(e\omega\log 2)$, e being the exponential. Hence each factor in $\prod_1$ is less than $1+1/(e\omega\log 2)$, so that

$$\log\prod{}_1 \leqslant \log\left(\frac{1}{e\omega\log 2}+1\right)\cdot\pi(\Omega)$$
$$\sim \log\log\log n\cdot\frac{(\log n)^{1-\eta}}{(1-\eta)\log\log n}$$
$$=o\left(\frac{\log n}{\log\log n}\right).$$

Now

$$\log d(n)=\omega\log n+\log\prod{}_1+\log\prod{}_2$$
$$<(1+\varepsilon)\log 2\cdot\frac{\log n}{\log\log n}+(\delta-\varepsilon)\log 2\cdot\frac{\log n}{\log\log n},\quad \text{for}\quad n\geqslant N(\delta),$$
$$=(1+\delta)\log 2\cdot\frac{\log n}{\log\log n}. \tag{83}$$

Now (83) and (81) give the required result.

Notes on Chapter I

As general references see the fundamental paper by A. Selberg, *Annals of Math.* 50 (1949), 305—313, as well as the papers by E. Wirsing, *Journal für Math.* 211 (1962), 205—214; 214/215 (1963), 1—18, and by E. Bombieri, *Riv. Mat. Univ. Parma* (2) 3 (1962), 393—440. See also A. Selberg, 11te Skandinaviske Matematikerkongress 1949, Oslo 1952, 13—22.

For classical proofs of the prime number theorem, see Ingham's *Tract*, Ch. II, and the author's *Introduction*, Ch. XI.

§ 1. For formulas (3) and (4) see, for instance, the author's *Introduction*, 56—58.

Formula (16) is due to T. Tatuzawa and K. Iseki, *Proc. Jap. Acad.* 27 (1951), 340—342.

If $M(x)=\sum_{n\leqslant x}\mu(n)$, $x\geqslant 1$, and we set $f(x)=M(x)$ in formula (4), we get $\sum_{n\leqslant x} f\left(\frac{x}{n}\right)=1$. If we now use formula (16) with $F(x)=f(x)$, we obtain

$$M(x)\log x+\sum_{n\leqslant x} M\left(\frac{x}{n}\right)\Lambda(n)=O(x),$$

since

$$\sum_{n\leqslant x} \mu(n)\log\frac{x}{n} = O(x),$$

as can be seen by partial summation from $M(x)=O(x)$. See Landau's *Vorlesungen*, II, 158.

For approximating a sum by an integral, as in (19), or vice versa, the standard device is a form of Abel's summation formula proved, for instance, in the author's *Introduction*, 78.

Let $0\leqslant\lambda_1\leqslant\lambda_2\leqslant\ldots$ be a sequence of real numbers, such that $\lambda_n\to\infty$ as $n\to\infty$, and let (a_n) be a sequence of complex numbers. Let $A(x)=\sum_{\lambda_n\leqslant x} a_n$, and $\varphi(x)$ a complex-valued function defined for $x\geqslant 0$. Then

$$\sum_{n=1}^{k} a_n\varphi(\lambda_n)=A(\lambda_k)\varphi(\lambda_k) - \sum_{n=1}^{k-1} A(\lambda_n)(\varphi(\lambda_{n+1})-\varphi(\lambda_n)).$$

If φ has a continuous derivative φ' in $(0,\infty)$, and $x\geqslant\lambda_1$, then

$$\sum_{\lambda_n\leqslant x} a_n\varphi(\lambda_n)=A(x)\varphi(x) - \int_{\lambda_1}^{x} A(t)\varphi'(t)\,dt.$$

If we set $a_n=1$ for all n, and $\lambda_n=k+n-1$, where k is a fixed, positive integer, then $A(x)=[x]-k+1$, and we have

$$\sum_{k\leqslant n\leqslant x} \varphi(n)=([x]-k+1)\varphi(x) - \int_{k}^{x}([t]-k+1)\varphi'(t)\,dt$$

$$=[x]\varphi(x)-(k-1)\varphi(k) - \int_{k}^{x}[t]\varphi'(t)\,dt.$$

Since

$$\int_{k}^{x} t\varphi'(t)\,dt=x\varphi(x)-k\varphi(k) - \int_{k}^{x}\varphi(t)\,dt,$$

we have

$$\sum_{k\leqslant n\leqslant x} \varphi(n) = \int_{k}^{x}\varphi(t)\,dt + \int_{k}^{x}(t-[t])\varphi'(t)\,dt+\varphi(k)-(x-[x])\varphi(x).$$

If φ' preserves the same sign throughout the interval $k\leqslant t\leqslant x$, then we have

$$\sum_{k\leqslant n\leqslant x} \varphi(n) = \int_{k}^{x}\varphi(t)\,dt+O(|\varphi(k)|+|\varphi(x)|).$$

If, in addition, $\varphi(t)$ is a non-negative, monotone decreasing function for $t\geqslant k$, with $\lim\limits_{t\to\infty}\varphi(t)=0$, then

$$\sum_{k\leqslant n\leqslant x} \varphi(n) = \int_{k}^{x}\varphi(t)\,dt+c+O(\varphi(x)),$$

where

$$c = \int_k^\infty (t-[t])\varphi'(t)\,dt + \varphi(k).$$

See the author's *Introduction*, 50.

Selberg's original proof of Theorem 1 is somewhat different, and makes no use of formula (16). See *Annals of Math.* 50 (1949), 305.

Since, by (7), we have

$$\sum_{n\leqslant x} \frac{\mu(n)}{n} = O(1), \quad \sum_{n\leqslant x} \frac{\mu(n)}{n}\log\frac{x}{n} = O(1), \quad \sum_{n\leqslant x} \frac{\mu(n)}{n}\log^2\frac{x}{n} = 2\log x + O(1),$$

Theorem 1 follows from the formula

$$\sum_{mn\leqslant x} \Lambda(m)\Lambda(n) + \sum_{n\leqslant x} \Lambda(n)\log n = \sum_{mn\leqslant x} \mu(n)\log^2 m\,, \qquad (*)$$

if we use the approximation

$$\sum_{m\leqslant \frac{x}{n}} \log^2 m = \frac{x}{n}\log^2\frac{x}{n} - 2\frac{x}{n}\left(\log\frac{x}{n} - 1\right) + O\left(\log^2\frac{x}{n}\right).$$

Formula (∗) can be obtained by comparing the coefficients of the product of two Dirichlet series obtained in two different ways (see the author's *Introduction*, 111—117).

If s is real, and (a_k), (b_j) are sequences of real numbers, and the Dirichlet series $\sum_{k=1}^{\infty} a_k k^{-s}$, $\sum_{j=1}^{\infty} b_j j^{-s}$ are absolutely convergent, for given s, and

$$f(s) = \sum_{k=1}^{\infty} a_k k^{-s}, \qquad g(s) = \sum_{j=1}^{\infty} b_j j^{-s},$$

then

$$f(s)\cdot g(s) = \sum_{n=1}^{\infty} c_n n^{-s}, \qquad c_n = \sum_{kj=n} a_k b_j.$$

If the summatory functions of the coefficients (a_k), (b_j), (c_n) are denoted by $A(x) = \sum_{k\leqslant x} a_k$, $B(x) = \sum_{j\leqslant x} b_j$, $C(x) = \sum_{n\leqslant x} c_n$, then

$$\sum_{\substack{kj\leqslant x\\ k>0,\, j>0}} a_k b_j = \sum_{0<k\leqslant y} a_k B\left(\frac{x}{k}\right) + \sum_{0<j\leqslant z} b_j A\left(\frac{x}{j}\right) - A(y)B(z), \qquad (**)$$

where $yz = x$. If we consider the rectangular hyperbola $yz = x$ for $x \geqslant 1$, and the set of lattice-points (k,j), such that $kj \leqslant x$, $k > 0$, $j > 0$, this set can be split up into two distinct subsets according to the conditions $0<k\leqslant y,\ 0<j\leqslant \frac{x}{k}$, and $y<k\leqslant x,\ j\leqslant \frac{x}{k}$. Hence

$$\sum_{kj\leqslant x} a_k b_j = \sum_{\substack{k\leqslant y\\ j\leqslant \frac{x}{k}}} a_k b_j + \sum_{\substack{y<k\leqslant x\\ j\leqslant \frac{x}{k}}} a_k b_j .$$

The last sum is, however, equal to

$$\sum_{\substack{j\leqslant z\\ k\leqslant \frac{x}{j}}} a_k b_j - \sum_{\substack{j\leqslant z\\ k\leqslant y}} a_k b_j ,$$

and formula (**) is a consequence.

Thus, if $s>1,\ a_k=1,\ b_j=\mu(j)$, then we have

$$\left(\sum_{k=1}^{\infty}\frac{1}{k^s}\right)\cdot\left(\sum_{j=1}^{\infty}\frac{\mu(j)}{j^s}\right)=1,$$

because of (2). If $\zeta(s)=\sum_{k=1}^{\infty}k^{-s},\ s>1$, then

$$\frac{1}{\zeta(s)}=\sum_{j=1}^{\infty}\frac{\mu(j)}{j^s},\qquad s>1,$$

and

$$\zeta^2(s)=\sum_{n=1}^{\infty}d(n)n^{-s},\qquad s>1,$$

since $\sum_{kj=n} 1=d(n)$, where $d(n)$ is the number of positive divisors of the positive integer n.

If we set $a_k=1,\ b_j=1,\ y=z=\sqrt{x}$, we get from (**),

$$\sum_{n\leqslant x} d(n)=x\log x+(2\gamma-1)x+O(\sqrt{x}),$$

because of formula (5).

If we note that

$$-\frac{\zeta'(s)}{\zeta(s)}=\sum_{n=1}^{\infty}\frac{\Lambda(n)}{n^s},\qquad s>1,$$

(as in the author's *Introduction*, 77) where the dash indicates differentiation, then the identity

$$\left(\frac{\zeta'(s)}{\zeta(s)}\right)^2+\left(\frac{\zeta'(s)}{\zeta(s)}\right)'=\frac{\zeta''(s)}{\zeta(s)}$$

gives formula (*) mentioned above. By partial summation of the second term in (*) we get Theorem 1, if we use the estimate $\psi(x)=O(x)$, which can be proved without the use of function theory (see the author's *Introduction*, 67).

The proof of (27) is in Selberg's paper, loc. cit. The deduction of the prime number theorem from (27) can be done in several ways. Selberg's original proof is given in the *Annals of Math.* 50 (1949), 305–313. There is a proof by P. Erdös, *Proc. Nat. Acad. Sci. USA* 35 (1949), 374—384. See also H. R. Pitt, *Tauberian theorems*, (Oxford) 1958, 160. A simple proof is given by V. Nevanlinna in *Commentationes Physico-Mathematicae* (Finland), XXVII, 3 (1962).

The prime number theorem is equivalent to the proposition that $p_n \sim n \log n$, where p_n denotes the nth prime (see the author's *Introduction*, 129—130). It is also known to be equivalent to either of the assertions $M(x)=o(x)$, or $g(x)=o(1)$, as $x\to\infty$, where $M(x)=\sum_{n\leqslant x}\mu(n)$, $g(x)=\sum_{n\leqslant x}\frac{\mu(n)}{n}$, the equivalence being provable by elementary arguments. See, for instance, G. H. Hardy's *Divergent series* (Oxford) 1949, 378—380.

Atle Selberg has also given an elementary proof of the prime number theorem for arithmetical progressions, namely

$$\pi(x;k,l) \sim \frac{1}{\varphi(k)} \cdot \frac{x}{\log x}, \quad \text{as} \quad x\to\infty,$$

where k is a positive integer, $(k,l)=1$, and φ is Euler's arithmetical function, and

$$\pi(x;k,l) = \sum_{\substack{p\leqslant x \\ p\equiv l \pmod{k}}} 1,$$

where p is a prime, so that $\pi(x;1,1)=\pi(x)$. See the *Canadian Journal of Math.* 2 (1950), 66–78.

§ 2. The proof of Theorem 2 is due to E. Wirsing, *Journal für Math.* 214/215 (1963), 1—18. Theorem 3, and in particular the idea of approximating ρ by a continuous, piece-wise linear function σ which is almost everywhere differentiable, is due to Wirsing.

§ 3. Theorem 4 may be looked upon as a sharper form of Schwarz's inequality. That the constant multiplying FG on the right-hand side of (59) is less than 1 is the essential point of the theorem. See Wirsing's paper II, loc. cit.

The existence of the integral $\int_0^t h(x)dx$ is a consequence of the hypotheses, by Hilbert's inequality: If $p>1$, $p'=p/(p-1)$, and

$$\int_0^\infty f^p(x)dx \leqslant F, \qquad \int_0^\infty g^{p'}(y)dy \leqslant G, \qquad f(x)\geqslant 0, \qquad g(y)\geqslant 0,$$

then

$$\int_0^\infty\int_0^\infty \frac{f(x)g(y)}{x+y}\,dx\,dy < \frac{\pi}{\sin(\pi/p)} F^{\frac{1}{p}} G^{\frac{1}{p'}},$$

unless $f\equiv 0$ or $g\equiv 0$. See G. H. Hardy, J. E. Littlewood, and G. Pólya, *Inequalities*, (Cambridge) 1934, Ch. IX.

That $h(x)$ is continuous follows from the square-integrability of f and g together with the fact that if $f\in L_2(-\infty,\infty)$, then the function

$$\omega_f(t)=\left(\int_{-\infty}^\infty |f(x+t)-f(x)|^2\,dx\right)^{\frac{1}{2}}$$

is continuous in t, and tends to zero as $t\to 0$. See, for instance, S. Bochner and K. Chandrasekharan, *Fourier transforms* (Princeton) 1949, 98—99.

To prove (67), we note that for $b=x_{n,\nu}$, $c=x_{n,\nu-1}$, $\delta>0$, $n\geqslant n_0(\delta)$, we have

$$\left|-c^{-1}\int_c^b h(x)dx+(c^{-1}-b^{-1})\int_0^b h(x)dx\right|$$

$$=\left|c^{-1}\int_0^c h(x)dx-b^{-1}\int_0^b h(x)dx\right| < 2\delta,$$

by (58), so that

$$\begin{aligned}\left|\int_c^b h(x)dx\right| &< (b-c)\left|b^{-1}\int_0^b h(x)dx\right| + 2\delta c\\ &< \delta(x_{n,\nu}-x_{n,\nu-1})+2\delta x_{n+1} \qquad \text{(by (58))}\\ &= \{\delta+2\delta\varepsilon^{-2}(1+\varepsilon)\}(x_{n,\nu}-x_{n,\nu-1})\\ &< \varepsilon(x_{n,\nu}-x_{n,\nu-1}),\end{aligned}$$

if we take $\delta=\varepsilon^3/5$.

§ 4. Both Wirsing and Bombieri (loc. cit) prove not only the prime number theorem, but the estimate

$$\psi(x)-x=O\left(\frac{x}{\log^\alpha x}\right), \quad \text{as} \quad x\to\infty,$$

for any fixed positive α, without the use of the theory of functions of a complex variable. Bombieri does this by generalizing Selberg's formula, and Wirsing by the application of his inequality (Theorem 4) to the iterated convolution σ_ν obtained from $\sigma_0 = \sigma$ by means of the formula

$$\sigma_{\nu+1}(\xi) = \frac{1}{\xi} \int_0^\xi \sigma_\nu(\xi - \eta) \sigma'_\nu(\eta) \, d\eta.$$

They also prove corresponding results for the prime number theorem for arithmetical progressions. Much stronger estimates of the *error term* $\psi(x) - x$ are obtained in Chapters 3 and 4 by using the theory of Riemann's zeta-function.

§ 5. For the elementary results on $d(n)$ stated in § 5 see, for instance, the author's *Introduction*, 47—54. For the proof of Theorem 6, see Landau's *Primzahlen*, 220. The theorem is due to S. Wigert, *Arkiv för matematik, astronomi och fysik*, Bd. 3, No. 18, (1906—1907), 9. The prime number theorem is *not* necessary for the proof. See S. Ramanujan, *Proc. London Math. Soc.* (2) 14 (1915), 347—409, or *Collected papers* (Cambridge) 1927, No. 15, for more precise results.

Chapter II

The zeta-function of Riemann

§ 1. The functional equation. If s is a complex number, with $s=\sigma+it$, where σ and t are real, and $i^2=-1$, the zeta-function of Riemann ζ is defined by the relation

$$\zeta(s)=\sum_{n=1}^{\infty} n^{-s}, \qquad \sigma>1. \tag{1}$$

The Dirichlet series on the right converges absolutely for $\sigma>1$, and uniformly in any half-plane $\sigma\geqslant 1+\delta>1$, where it defines ζ as a regular analytic function. By *Euler's identity*, we can represent $\zeta(s)$, for $\sigma>1$, by an absolutely convergent infinite product, namely

$$\zeta(s)=\prod_p (1-p^{-s})^{-1}, \qquad \sigma>1, \tag{2}$$

where the product is taken over all primes p. It follows that $\zeta(s)\neq 0$ for $\sigma>1$.

We shall prove that ζ is regular all over the complex s-plane, except for a simple pole at $s=1$, with residue 1, and that it satisfies the *functional equation*

$$\pi^{-\frac{1}{2}s}\Gamma(\tfrac{1}{2}s)\zeta(s)=\pi^{-\frac{1}{2}(1-s)}\Gamma(\tfrac{1}{2}-\tfrac{1}{2}s)\zeta(1-s). \tag{3}$$

We shall make use of a form of *Poisson's summation formula* (see (4)) given by the following

THEOREM 1. *If* (i) *the function f is continuous in the interval* $-\infty<x<\infty$, (ii) *the series* $\sum_{k=-\infty}^{\infty} f(x+k)$ *converges uniformly in the interval* $-\frac{1}{2}\leqslant x\leqslant\frac{1}{2}$, *and* (iii) *the series* $\sum_{j=-\infty}^{\infty}\varphi(j)$ *converges, where* $\varphi(\alpha)$, *for any real* α, *is defined by*

$$\varphi(\alpha)=\int_{-\infty}^{\infty} f(x)e^{2\pi i\alpha x}\,dx=\lim_{T\to\infty}\int_{-T}^{T} f(x)e^{2\pi i\alpha x}\,dx,$$

the limit existing, then we have

$$\sum_{k=-\infty}^{\infty} f(k) = \sum_{j=-\infty}^{\infty} \varphi(j). \tag{4}$$

PROOF (BOCHNER). If j and n are positive integers, we have

$$\int_{-n-\frac{1}{2}}^{n+\frac{1}{2}} f(x) e^{2\pi ijx} dx = \int_{-\frac{1}{2}}^{\frac{1}{2}} \sum_{k=-n}^{n} f(x+k) e^{2\pi ijx} dx. \tag{5}$$

If $\lim_{n\to\infty} \sum_{k=-n}^{n} f(x+k) = g(x)$, the limit being uniform in $-\frac{1}{2} \leqslant x \leqslant \frac{1}{2}$, then g is continuous, and periodic with period 1. On letting $n \to \infty$ in (5), we have

$$\varphi(j) = \int_{-\frac{1}{2}}^{\frac{1}{2}} g(x) e^{2\pi ijx} dx,$$

the *Fourier coefficient* of g. By Fejér's theorem, the Fourier series of g is summable $(C, 1)$, at every point of continuity of g, to a sum which is equal to the value of g at that point. Thus, for $x=0$, we have

$$g(0) = \sum_{(C,1)} \varphi(j).$$

If we assume, however, that $\sum_{j=-\infty}^{\infty} \varphi(j)$ converges, then $\sum_{(C,1)} \varphi(j) = \sum_{j=-\infty}^{\infty} \varphi(j)$. Hence

$$\lim_{n\to\infty} \sum_{k=-n}^{n} f(k) = \lim_{m\to\infty} \sum_{j=-m}^{m} \varphi(j),$$

which proves the theorem.

We shall use Theorem 1 to prove

THEOREM 2 (RIEMANN). *The function* ζ, *defined by* (1), *is regular all over the complex s-plane, except for a simple pole at* $s=1$ *with residue* 1, *and satisfies the functional equation*

$$\pi^{-\frac{1}{2}s} \Gamma(\tfrac{1}{2}s) \zeta(s) = \pi^{-\frac{1}{2}(1-s)} \Gamma(\tfrac{1}{2} - \tfrac{1}{2}s) \zeta(1-s). \tag{3}$$

PROOF. If we set $f(x) = e^{-\pi x^2 t}$, $t>0$, in Theorem 1, and observe that $\varphi(\alpha) = t^{-\frac{1}{2}} e^{-\pi\alpha^2/t}$, then we have the *theta-relation*

$$\sum_{k=-\infty}^{\infty} e^{-\pi k^2 t} = \frac{1}{\sqrt{t}} \sum_{k=-\infty}^{\infty} e^{-\pi k^2/t}, \quad t>0. \tag{6}$$

This holds, by analytic continuation, also for complex t with $\operatorname{Re} t > 0$. If we write

$$\theta(t) = \sum_{k=1}^{\infty} e^{-\pi k^2 t}, \quad \operatorname{Re} t > 0, \tag{7}$$

then from (6) we have

$$\theta(t) = \frac{1}{2\sqrt{t}}\left(2\,\theta\left(\frac{1}{t}\right) + 1\right) - \frac{1}{2}. \tag{8}$$

We shall use this formula to establish the analytic continuation as well as the functional equation of ζ.

From the well-known representation of the gamma-function as an integral, we have

$$\Gamma(\tfrac{1}{2}s) = \int_0^{\infty} e^{-x} x^{s/2-1}\, dx, \qquad \sigma = \operatorname{Re} s > 0.$$

If n is a positive integer, we have, on writing $\pi n^2 x$ for x,

$$\Gamma(\tfrac{1}{2}s) = \int_0^{\infty} e^{-\pi n^2 x} (\pi n^2 x)^{s/2-1} \pi n^2\, dx = \pi^{s/2} n^s \int_0^{\infty} e^{-\pi n^2 x} x^{s/2-1}\, dx, \qquad \sigma > 0,$$

or

$$\frac{1}{n^s} = \frac{\pi^{s/2}}{\Gamma(\frac{1}{2}s)} \int_0^{\infty} e^{-\pi n^2 x} x^{s/2-1}\, dx, \qquad \sigma > 0.$$

Now, if $\sigma > 1$, then we have

$$\zeta(s) = \sum_{n=1}^{\infty} \frac{1}{n^s} = \frac{\pi^{s/2}}{\Gamma(\frac{1}{2}s)} \sum_{n=1}^{\infty} \int_0^{\infty} e^{-\pi n^2 x} x^{s/2-1}\, dx,$$

and since the series

$$\sum_{n=1}^{\infty} \int_0^{\infty} |e^{-\pi n^2 x} x^{s/2-1}|\, dx = \sum_{n=1}^{\infty} \int_0^{\infty} e^{-\pi n^2 x} x^{\sigma/2-1}\, dx = \sum_{n=1}^{\infty} \frac{\Gamma(\frac{1}{2}\sigma)}{\pi^{\frac{1}{2}\sigma} n^{\sigma}}$$

is convergent for $\sigma > 1$, we can interchange the order of summation and integration, and obtain

$$\zeta(s) = \frac{\pi^{s/2}}{\Gamma(\frac{1}{2}s)} \int_0^{\infty} \theta(x) x^{s/2-1}\, dx.$$

This can be rewritten as

$$\pi^{-\frac{1}{2}s}\Gamma(\tfrac{1}{2}s)\zeta(s) = \int_0^1 x^{s/2-1}\theta(x)\,dx + \int_1^\infty x^{s/2-1}\theta(x)\,dx$$

$$= \int_0^1 x^{s/2-1}\left(x^{-\frac{1}{2}}\theta\left(\frac{1}{x}\right) + \frac{1}{2\sqrt{x}} - \frac{1}{2}\right)dx + \int_1^\infty x^{s/2-1}\theta(x)\,dx$$

(by (8))

$$= \frac{1}{s-1} - \frac{1}{s} + \int_0^1 x^{s/2-\frac{3}{2}}\theta\left(\frac{1}{x}\right)dx + \int_1^\infty x^{s/2-1}\theta(x)\,dx$$

$$= \frac{1}{s(s-1)} + \int_1^\infty (x^{-s/2-\frac{1}{2}} + x^{s/2-1})\theta(x)\,dx. \tag{9}$$

The integral on the right-hand side of (9) converges uniformly in $-\infty < a \leqslant \sigma \leqslant b < +\infty$, for if $x \geqslant 1$, we have

$$|x^{-s/2-\frac{1}{2}} + x^{s/2-1}| \leqslant x^{-a/2-\frac{1}{2}} + x^{b/2-1},$$

while

$$\theta(x) < \sum_{n=1}^\infty e^{-\pi n x} = \frac{1}{e^{\pi x}-1}, \qquad x>0.$$

Hence the integral in (9) represents an entire function of s. Therefore

$$\pi^{-\frac{1}{2}s}\Gamma(\tfrac{1}{2}s)\zeta(s) - \frac{1}{s(s-1)}$$

is an entire function of s. Since $\dfrac{\pi^{\frac{1}{2}s}}{\Gamma(\frac{1}{2}s)}$ is entire, it follows that $\zeta(s) - \dfrac{1}{s(s-1)}\dfrac{\pi^{\frac{1}{2}s}}{\Gamma(\frac{1}{2}s)}$ is entire. But $s\Gamma(\frac{1}{2}s) = 2\Gamma(\frac{1}{2}s+1)$, so that $\zeta(s) - \dfrac{1}{s-1}\dfrac{\pi^{\frac{1}{2}s}}{2\Gamma(\frac{1}{2}s+1)}$ is entire. Since $\dfrac{\pi^{\frac{1}{2}}}{2\Gamma(\frac{3}{2})} = 1$, it follows that $\zeta(s) - \dfrac{1}{s-1}$ is entire.

Formula (9) thus provides the analytic continuation of ζ all over the s-plane, and establishes the functional equation at the same time. This completes the proof of Theorem 2.

If we define

$$\eta(s)=\pi^{-\frac{1}{2}s}\Gamma(\tfrac{1}{2}s)\,\zeta(s), \tag{10}$$

and

$$\xi(s)=\tfrac{1}{2}s(s-1)\eta(s), \tag{11}$$

we obtain, as an immediate consequence of Theorem 2, the following

THEOREM 3. (i) *The function* ξ *is entire, and* $\xi(s)=\xi(1-s)$; (ii) $\xi(s)$ *is real for* $t=0$, *and for* $\sigma=\frac{1}{2}$; (iii) $\xi(0)=\xi(1)=\frac{1}{2}$.

PROOF. Part (i) is a consequence of Theorem 2. Since ξ is clearly regular for $\sigma>0$, by the functional equation of ζ, it is also regular for $\sigma<1$, hence entire.

Since $\xi(s)$ is obviously real for real s, it takes conjugate values at conjugate points. Thus $\xi(\frac{1}{2}+it)$ and $\xi(\frac{1}{2}-it)$ are conjugates. They are equal by the functional equation. So they are real, which proves part (ii) of the theorem.

To prove part (iii) one has only to observe that $\xi(s)=\pi^{-\frac{1}{2}s}\Gamma(\frac{1}{2}s+1)\cdot(s-1)\zeta(s)$, so that $\xi(1)=\pi^{-\frac{1}{2}}\Gamma(\frac{3}{2})=\frac{1}{2}$, and by the functional equation, $\xi(0)=\frac{1}{2}$.

We shall now prove a result on the location of the zeros of ξ, assuming that it *has* zeros.

THEOREM 4. (i) *The zeros of* ξ, *if any, are all located in the strip* $0\leqslant\sigma\leqslant 1$, *and lie symmetrically about the lines* $t=0$, *and* $\sigma=\frac{1}{2}$;

(ii) *the zeros of* ζ *are identical, in position and multiplicity, with those of* ξ, *except that* ζ *has a simple zero at each of the points* $s=-2,-4,-6,\ldots$;

(iii) ξ *has no zeros on the line* $t=0$.

PROOF. If we define $H(s)$ by the relation

$$\xi(s)=\pi^{-\frac{1}{2}s}(s-1)\Gamma(\tfrac{1}{2}s+1)\zeta(s)=H(s)\zeta(s), \tag{12}$$

then $H(s)\neq 0$ for $\sigma>1$, and $\zeta(s)\neq 0$ for $\sigma>1$ by (2). Hence $\xi(s)\neq 0$ for $\sigma>1$, and by the functional equation, $\xi(s)\neq 0$ for $\sigma<0$. The zeros of ξ, if any, are symmetrical about the real axis, since $\xi(\sigma+it)$ and $\xi(\sigma-it)$ are conjugates, and symmetrical about the point $s=\frac{1}{2}$, since $\xi(s)=\xi(1-s)$. They are therefore symmetrical about the line $\sigma=\frac{1}{2}$. Thus part (i) is proved.

The zeros of ζ differ from those of ξ only in so far as H has zeros or poles. The only zero of H is at $s=1$. But this is not a zero of ξ, since $\xi(1)=\frac{1}{2}$ by Theorem 3, nor of ζ, for it is a pole of the latter.

The poles of H are simple, at $s=-2,-4,-6,\ldots$. At these points ξ is regular, and not zero, by (i); hence they are simple zeros of ζ, which proves part (ii).

To prove part (iii), we have only to show, in view of (i) and (ii), and the fact that $\xi(0)=\xi(1)=\frac{1}{2}$, which was proved in Theorem 3, that $\zeta(\sigma)\neq 0$ for $0<\sigma<1$. This follows from the fact that, for $\sigma>0$, we have

$$(1-2^{1-s})\zeta(s)=(1-2^{-s})+(3^{-s}-4^{-s})+\cdots. \tag{13}$$

This is obvious for $\sigma>1$, and since both sides are regular for $\sigma>0$, it holds also for $\sigma>0$. The left-hand side is regular for $\sigma>0$, after Theorem 2. So is the right-hand side, since

$$\left|\frac{1}{(2n-1)^s}-\frac{1}{(2n)^s}\right| = \left|s\int_{2n-1}^{2n}\frac{dx}{x^{s+1}}\right|$$
$$\leqslant \frac{|s|}{(2n-1)^{\sigma+1}} \leqslant \frac{\Delta}{(2n-1)^{\delta+1}},$$

if $\sigma>\delta$, and $|s|<\Delta$, for fixed positive numbers δ, Δ. Relation (13) implies that, for $0<\sigma<1$, we have $(1-2^{1-\sigma})\zeta(\sigma)>0$, or $\zeta(\sigma)<0$, which proves part (iii).

REMARKS. The strip $0\leqslant\sigma\leqslant 1$ is called the *critical strip*, and the line $\sigma=\frac{1}{2}$ the *critical line*. The points $s=-2,-4,-6,\ldots$ are called the *trivial zeros of* ζ.

§ 2. The Riemann-von Mangoldt formula. We are yet to prove that ξ has zeros, or that ζ has non-real zeros. We shall do this by obtaining an estimate for the number $N(T)$ of non-real zeros of ζ in the rectangle $0\leqslant\sigma\leqslant 1$, $0\leqslant t\leqslant T$, which will show in fact that $N(T)\to\infty$ as $T\to\infty$. For this purpose we require some estimates for $|\zeta(s)|$.

THEOREM 5. *If* $\sigma\geqslant 1$, $t\geqslant 2$, *we have*

$$|\zeta(s)|<c\log t, \tag{14}$$

where c is a positive constant. On the other hand, if δ *is such that* $0<\delta<1$, *and* $\sigma\geqslant\delta$, $t\geqslant 1$, *then*

$$|\zeta(s)|<c(\delta)t^{1-\delta}, \tag{15}$$

where $c(\delta)$ *is a positive constant depending only on* δ.

PROOF. In any fixed half-plane $\sigma\geqslant 1+\varepsilon>1$, $\zeta(s)$ is bounded, for

$$|\zeta(s)|\leqslant\zeta(\sigma)\leqslant\zeta(1+\varepsilon).$$

By Abel's summation formula, we have

$$\sum_{n\leqslant x}\frac{1}{n^s}=\frac{[x]}{x^s}+s\int_1^x\frac{[u]}{u^{s+1}}\,du\,,\qquad x\geqslant 1,\ \sigma>0\,,\tag{16}$$

where, as usual, $[x]$ denotes the integral part of x. This implies that

$$\zeta(s)=\sum_{n=1}^{\infty}\frac{1}{n^s}=\frac{s}{s-1}-s\int_1^{\infty}\frac{x-[x]}{x^{s+1}}\,dx\,,\quad\text{for}\quad \sigma>1\,.\tag{17}$$

Since

$$\left|\frac{x-[x]}{x^{s+1}}\right|<\frac{1}{x^{\sigma+1}}\,,$$

the last integral in (17) converges uniformly for $\sigma>\delta$, where δ is any fixed positive number, and therefore represents a regular function of s for $\sigma>0$. Relation (17) gives the analytic continuation of ζ for $\sigma>0$, a fact already established in Theorem 2. From (16) and (17) we have, for $\sigma>0$, $x\geqslant 1$,

$$\zeta(s)-\sum_{n\leqslant x}\frac{1}{n^s}=-s\int_x^{\infty}\frac{u-[u]}{u^{s+1}}\,du+\frac{1}{(s-1)x^{s-1}}+\frac{x-[x]}{x^s}\,.$$

Hence, if $t>0$,

$$\begin{aligned}|\zeta(s)|&<\sum_{n\leqslant x}\frac{1}{n^\sigma}+\frac{1}{t\,x^{\sigma-1}}+\frac{1}{x^\sigma}+|s|\int_x^{\infty}\frac{du}{u^{\sigma+1}}\\&<\sum_{n\leqslant x}\frac{1}{n^\sigma}+\frac{1}{t\,x^{\sigma-1}}+\frac{1}{x^\sigma}+\left(\frac{\sigma+t}{\sigma}\right)\frac{1}{x^\sigma}\,.\end{aligned}\tag{18}$$

If $\sigma\geqslant 1$, $t\geqslant 1$, $x\geqslant 1$, we have

$$\begin{aligned}|\zeta(s)|&<\sum_{n\leqslant x}\frac{1}{n}+\frac{1}{t}+\frac{1}{x}+\frac{1+t}{x}\\&\leqslant(\log x+1)+3+\frac{t}{x}\,.\end{aligned}$$

Taking $t=x$, we obtain

$$|\zeta(s)|<c\log t\,,\qquad \sigma\geqslant 1,\ t\geqslant 2\,,$$

which is (14). On the other hand, if $\sigma \geqslant \delta, 0<\delta<1, t \geqslant 1, x \geqslant 1,$ then from (18) we have

$$|\zeta(s)| < \sum_{n \leqslant x} \frac{1}{n^\sigma} + \frac{1}{t x^{\delta-1}} + \left(2 + \frac{t}{\delta}\right) \frac{1}{x^\delta}$$

$$< \int_0^{[x]} \frac{du}{u^\delta} + \frac{x^{1-\delta}}{t} + \frac{3t}{\delta x^\delta}$$

$$\leqslant \frac{x^{1-\delta}}{1-\delta} + x^{1-\delta} + \frac{3t}{\delta x^\delta}.$$

Taking $x=t$, we obtain (15).

Apart from the estimates for $|\zeta(s)|$ given by Theorem 5, we need the following function-theoretical

LEMMA A. *If* $R>0$, *and* f *is a function regular for* $|z-z_0| \leqslant R$, *and has (at least)* n *zeros in* $|z-z_0| \leqslant r<R$, *multiple zeros being counted according to their order of multiplicity, then, if* $f(z_0) \neq 0$, *we have*

$$\left(\frac{R}{r}\right)^n \leqslant \frac{M}{|f(z_0)|}, \tag{19}$$

where $M=\max |f(z)|$ *for* $|z-z_0|=R$.

PROOF. We may clearly suppose that $z_0=0$, for otherwise we can make the substitution $z=z_0+z'$, and deal with the new variable z'. If $a_1, a_2, \ldots, a_n$ are the zeros of f in $|z| \leqslant r$, each zero being counted according to its multiplicity, then

$$f(z)=\varphi(z) \prod_{\nu=1}^{n} \frac{R(z-a_\nu)}{R^2-\bar{a}_\nu z},$$

where $\bar{a}_\nu$ is the complex conjugate of a_ν, and φ is regular in $|z| \leqslant R$. Each factor in the above product has modulus 1 for $|z|=R$. Hence

$$|\varphi(z)|=|f(z)| \leqslant M, \, |z|=R,$$

which implies, by the maximum modulus principle, that $|\varphi(0)| \leqslant M$, or

$$|f(0)|=|\varphi(0)| \prod_{\nu=1}^{n} \frac{|a_\nu|}{R} \leqslant M\left(\frac{r}{R}\right)^n,$$

which proves (19), since $|f(0)| \neq 0$.

We are now in a position to estimate $N(T)$, defined for $T>0$ as the number of zeros of ζ in the rectangle $0 \leqslant \sigma \leqslant 1, 0 \leqslant t \leqslant T$. By Theorem 4, $N(T)$ is the number of zeros $\rho=\beta+i\gamma$ of ζ (where β and γ are real, $i^2=-1$) for which $0<\gamma \leqslant T$.

THEOREM 6 (RIEMANN-VON MANGOLDT). *We have*

$$N(T)=\frac{T}{2\pi}\log\frac{T}{2\pi}-\frac{T}{2\pi}+O(\log T), \tag{20}$$

as $T\to\infty$.

PROOF (BACKLUND). Suppose that T is not equal to *any* γ, and that $T>3$. Consider the rectangle $\mathscr{R}$ with vertices $2+iT$, $-1+iT$, $-1-iT$, and $2-iT$, in that order. The function ξ has $2N(T)$ zeros in the interior of $\mathscr{R}$, and none on the boundary. Hence

$$N(T)=\frac{1}{4\pi}\operatorname{Im}\left(\int_{\mathscr{R}}\frac{\xi'(s)}{\xi(s)}\,ds\right). \tag{21}$$

By (10) and (11),

$$\frac{\xi'(s)}{\xi(s)}=\frac{1}{s}+\frac{1}{s-1}+\frac{\eta'(s)}{\eta(s)},\qquad \eta(s)=\pi^{-s/2}\Gamma(\tfrac{1}{2}s)\zeta(s).$$

Obviously

$$\operatorname{Im}\left(\int_{\mathscr{R}}\left(\frac{1}{s}+\frac{1}{s-1}\right)ds\right)=4\pi.$$

Since $\eta(s)=\eta(1-s)$, and $\eta(\sigma\pm it)$ are conjugates, we have

$$\operatorname{Im}\left(\int_{\mathscr{R}}\frac{\eta'(s)}{\eta(s)}\,ds\right)=4\operatorname{Im}\left(\int_{\mathscr{L}}\frac{\eta'(s)}{\eta(s)}\,ds\right),$$

where $\mathscr{L}$ is that part of the boundary of $\mathscr{R}$ which runs from the point $s=2$ to the points $s=2+iT$ (say $\mathscr{L}_1$), together with that part which runs from $s=2+iT$ to the point $s=\frac{1}{2}+iT$ (say $\mathscr{L}_2$). Now

$$\begin{aligned}\operatorname{Im}\left(\int_{\mathscr{L}}\frac{\eta'(s)}{\eta(s)}\,ds\right)&=\operatorname{Im}\left(\int_{\mathscr{L}}(-\tfrac{1}{2}\log\pi)\,ds+\int_{\mathscr{L}}\frac{\frac{1}{2}\Gamma'(\frac{1}{2}s)}{\Gamma(\frac{1}{2}s)}\,ds+\int_{\mathscr{L}}\frac{\zeta'(s)}{\zeta(s)}\,ds\right)\\&=-\tfrac{1}{2}\log\pi\cdot T+\operatorname{Im}\left(\int_{\mathscr{L}}\frac{\frac{1}{2}\Gamma'(\frac{1}{2}s)}{\Gamma(\frac{1}{2}s)}\,ds+\int_{\mathscr{L}}\frac{\zeta'(s)}{\zeta(s)}\,ds\right).\end{aligned} \tag{22}$$

Since

$$\operatorname{Im}\left(\int_{\mathscr{L}}\frac{\frac{1}{2}\Gamma'(\frac{1}{2}s)}{\Gamma(\frac{1}{2}s)}\,ds\right)=\operatorname{Im}\log\Gamma(\tfrac{1}{4}+\tfrac{1}{2}iT)-\operatorname{Im}\log\Gamma(1),$$

and

$$\log\Gamma(z+\alpha)=(z+\alpha-\tfrac{1}{2})\log z-z+\tfrac{1}{2}\log 2\pi+O\left(\frac{1}{|z|}\right), \tag{23}$$

as $|z|\to\infty$, uniformly in any angle $|\arg z|\leqslant\pi-\varepsilon<\pi$, if α is bounded, it follows that

$$\operatorname{Im}\left(\int_{\mathscr{L}} \frac{\frac{1}{2}\Gamma'(\frac{1}{2}s)}{\Gamma(\frac{1}{2}s)}\,ds\right) = \operatorname{Im}((-\tfrac{1}{4}+\tfrac{1}{2}iT)\log(\tfrac{1}{2}iT)-\tfrac{1}{2}iT+\tfrac{1}{2}\log 2\pi)+O\left(\frac{1}{T}\right)$$

$$= \tfrac{1}{2}T\log(\tfrac{1}{2}T)-\tfrac{1}{2}T-\tfrac{1}{8}\pi+O\left(\frac{1}{T}\right),$$

as $T\to\infty$. Collecting the results, we have

$$N(T) = \frac{1}{4\pi}\operatorname{Im}\left(\int_{\mathscr{R}} \frac{\xi'(s)}{\xi(s)}\,ds\right)$$

$$= \frac{T}{2\pi}\log\frac{T}{2\pi} - \frac{T}{2\pi} + \frac{7}{8} + \frac{1}{\pi}\operatorname{Im}\left(\int_{\mathscr{L}} \frac{\zeta'(s)}{\zeta(s)}\,ds\right) + O\left(\frac{1}{T}\right). \qquad (24)$$

Now let m be the number of distinct points s' on $\mathscr{L}$, exclusive of the end-points, at which $\operatorname{Re}\zeta(s')=0$. This number is necessarily finite, as we shall presently see. On any of the $m+1$ segments of $\mathscr{L}$ into which $\mathscr{L}$ is divided by these points, $\operatorname{Re}\zeta(s)$ cannot change sign. Let S denote one such segment. Then

$$\operatorname{Im}\int_{S} \frac{\zeta'(s)}{\zeta(s)}\,ds = \operatorname{Im}\int_{\zeta(S)} \frac{dz}{z},$$

where $\zeta(S)$ is the image of S under the map $s\to\zeta(s)$. Clearly $\zeta(S)$ is contained in one of the half-planes $\operatorname{Re}z\geqslant 0$, or $\operatorname{Re}z\leqslant 0$. If the curve $\zeta(S)$ begins at the point $z=a$ and ends at the point $z=b$, and $\overline{a,b}$ denotes the line segment joining a and b, then we have, by Cauchy's theorem,

$$\int_{\zeta(S)} \frac{dz}{z} = \int_{K_{a,b}} \frac{dz}{z},$$

where $K_{a,b}$ denotes the semi-circle, with $\overline{a,b}$ as base which lies in the same half-plane as $\zeta(S)$. Hence

$$\left|\operatorname{Im}\int_{\zeta(S)} \frac{\zeta'(s)}{\zeta(s)}\,ds\right| = \left|\operatorname{Im}\int_{K_{a,b}} \frac{dz}{z}\right| \leqslant \pi.$$

It follows that

$$\left|\operatorname{Im}\int_{\mathscr{L}} \frac{\zeta'(s)}{\zeta(s)}\,ds\right| \leqslant (m+1)\pi.$$

Now no point s' can be on $\mathscr{L}_1$, since

$$\operatorname{Re}\zeta(2+it) \geqslant 1 - \sum_{n=2}^{\infty} \frac{1}{n^2} > 1 - \frac{1}{2^2} - \int_2^{\infty} \frac{du}{u^2} = \frac{1}{4}. \tag{25}$$

Thus m is the number of distinct points σ in the interval $\frac{1}{2}<\sigma<2$, at which $\operatorname{Re}\zeta(\sigma+iT)=0$. It is equal to the number of distinct zeros of $g(s)=\frac{1}{2}(\zeta(s+iT)+\zeta(s-iT))$ on the real axis, in the interval $\frac{1}{2}<\sigma<2$, for $g(\sigma)=\operatorname{Re}\zeta(\sigma+iT)$ for real σ, because $\zeta(\sigma\pm iT)$ are conjugates. Since g is regular except for $s=1\pm iT$, and not identically zero, m must be finite. We shall find an upper bound for m, with the aid of the estimates for $|\zeta(s)|$ obtained in Theorem 5, and by using the lemma proved above.

We shall apply the lemma to the function $g(s)=\frac{1}{2}(\zeta(s+iT)+\zeta(s-iT))$, and to the two circles: $|s-2|\leqslant\frac{7}{4}$, and $|s-2|\leqslant\frac{3}{2}$. Since we have supposed that $T>3$, the function g is regular in the larger circle. Further $g(2)\neq 0$, since actually $g(2)=\operatorname{Re}(\zeta(2+iT))>\frac{1}{4}$, by (25). If we use the estimate for $|\zeta(s)|$ obtained in (15), with $\delta=\frac{1}{4}$, we get, for every s such that $|s-2|=\frac{7}{4}$, the inequality

$$|g(s)|<\tfrac{1}{2}\cdot c(|t+T|^{\frac{3}{4}}+|t-T|^{\frac{3}{4}})<c_1(2+T)^{\frac{3}{4}},$$

since $\sigma\geqslant\frac{1}{4}$, and $1\leqslant|t\pm T|<2+T$. Hence by an application of (19)

$$\left(\frac{7}{6}\right)^m < \frac{c_1(2+T)^{\frac{3}{4}}}{\left(\frac{1}{4}\right)} < T, \quad \text{for} \quad T>T_0\geqslant 3.$$

Therefore $m<c_2\log T$, for $T>T_0$, for a positive constant c_2. Substituting this into the inequality

$$\left|\operatorname{Im}\int_{\mathscr{L}}\frac{\zeta'(s)}{\zeta(s)}\,ds\right|\leqslant(m+1)\pi,$$

and combining it with (24), we deduce that

$$N(T)=\frac{T}{2\pi}\log\frac{T}{2\pi}-\frac{T}{2\pi}+O(\log T),$$

as $T\to\infty$, on the assumption that T is not equal to any γ. This restriction does not affect the result, since we can, if necessary, replace T by a larger T', different from any γ, and let $T'\to T+0$. The proof of Theorem 6 is therefore complete.

COROLLARY 1. *For any fixed* $h>0$, *we have*

$$N(T+h)-N(T)=O(\log T), \tag{26}$$

as $T\to\infty$.

PROOF. If we write $f(t)=\left(\frac{t}{2\pi}\right)\log\left(\frac{t}{2\pi}\right)-\frac{t}{2\pi}$, then $f(t+h)-f(t)$ $=hf'(t+\alpha h)$, for $0<\alpha<1$, where $f'(t)=\left(\frac{1}{2\pi}\right)\log\left(\frac{t}{2\pi}\right)$. Combining this remark with Theorem 6, we obtain (26).

COROLLARY 2. *If* ρ *denotes a non-real zero of* ζ, *and* $\rho=\beta+i\gamma$, *where* β, γ *are real, and* $i^2=-1$, *then*

$$S=\sum_{0<\gamma\leqslant T}\frac{1}{\gamma}=O(\log^2 T), \tag{27}$$

and

$$S'=\sum_{\gamma>T}\frac{1}{\gamma^2}=O\left(\frac{\log T}{T}\right). \tag{28}$$

PROOF. Let

$$s_m=\sum\frac{1}{\gamma}, \qquad m<\gamma\leqslant m+1,$$

and

$$s'_m=\sum\frac{1}{\gamma^2}, \qquad m<\gamma\leqslant m+1.$$

Then

$$S\leqslant\sum_{m=0}^{[T]} s_m, \qquad S'\leqslant\sum_{m=[T]}^{\infty} s'_m.$$

If $m\geqslant 1$, the number of terms in s_m, or in s'_m, is $N(m+1)-N(m)=v_m$, say. By (26) we know that $v_m=O(\log m)$. Further $s_m\leqslant v_m/m$, $s'_m\leqslant v_m/m^2$. Hence, as $T\to\infty$, we have

$$S=O(1)+O\left(\sum_{m=2}^{[T]}\frac{\log m}{m}\right)=O(\log^2 T),$$

and

$$S'=O\left(\sum_{[T]}^{\infty}\frac{\log m}{m^2}\right)=O\left(\frac{\log T}{T}\right).$$

COROLLARY 3. *If the non-real zeros of* ζ, *with positive imaginary part, are arranged as a sequence* $\rho_n=\beta_n+i\gamma_n$, $\gamma_{n+1}\geqslant\gamma_n$, *then*

$$|\rho_n|\sim\gamma_n\sim\frac{2\pi n}{\log n}, \quad \textit{as} \quad n\to\infty. \tag{29}$$

PROOF. Since $N(\gamma_n - 1) < n \leqslant N(\gamma_n + 1)$, and

$$2\pi N(\gamma_n \pm 1) \sim (\gamma_n \pm 1)\log(\gamma_n \pm 1) \sim \gamma_n \log \gamma_n,$$

we have $2\pi n \sim \gamma_n \log \gamma_n$, or $\gamma_n \sim \dfrac{2\pi n}{\log \gamma_n}$, whence $\log n \sim \log \gamma_n$, so that

$$\gamma_n \sim \frac{2\pi n}{\log \gamma_n} \sim \frac{2\pi n}{\log n}.$$

Further $\gamma_n \leqslant |\rho_n| < \gamma_n + 1$, which implies that $|\rho_n| \sim \gamma_n$.

REMARK. It follows that

$$\sum_\rho \frac{1}{|\rho|(1+\log|\rho|)^\alpha} \tag{30}$$

converges for $\alpha > 2$, and diverges for $\alpha \leqslant 2$.

§ 3. The entire function ξ. We have proved in Theorem 3 that ξ is an entire function. We shall now prove that it is of *order* 1.

The order ω of the entire function ξ is defined by

$$\omega = \limsup_{R\to\infty} \frac{\log\log M(R)}{\log R},$$

where $M(R) = \max|\xi(s)|$ for $|s| = R > 0$. We shall prove that

$$\log M(R) \sim \tfrac{1}{2} R \log R, \quad \text{as} \quad R \to \infty, \tag{31}$$

with the help of (15) and (23).

If $\sigma \geqslant 2$, we have $|\zeta(s)| \leqslant \zeta(2)$; and for $\sigma \geqslant \frac{1}{2}$, $t \geqslant 2$, we have from (15), $|\zeta(s)| < c_1 |t|^{\frac{1}{2}}$, which implies that $|\zeta(s)| < c_2 |s|^{\frac{1}{2}}$ for $\sigma \geqslant \frac{1}{2}$, $|s| > 3$. If we further use (23), and (10) and (11), we get for $\sigma \geqslant \frac{1}{2}$, $|s| = R > 3$,

$$|\xi(s)| < e^{|\frac{1}{2}s\log\frac{1}{2}s| + c_3|s|} < e^{\frac{1}{2}R\log R + c_4 R}, \tag{32}$$

since $\log\frac{1}{2}s = -\log 2 + \log|s| + i\arg s$, and $|\arg s| < \frac{1}{2}\pi$. Since $\xi(s) = \xi(1-s)$, we infer that

$$|\xi(s)| < e^{\frac{1}{2}|1-s|\cdot\log|1-s| + c_4|1-s|} < e^{\frac{1}{2}R\log R + c_5 R} \tag{33}$$

for $\sigma \leqslant \frac{1}{2}$, $|s| = R > 4$. Combining (32) and (33), we have

$$|\xi(s)| < e^{\frac{1}{2}R\log R + c_6 R}, \qquad |s| = R > 4. \tag{34}$$

On the other hand, if $R > 2$, we have

$$M(R) \geqslant \xi(R) > 1 \cdot \pi^{-R/2}\Gamma(\tfrac{1}{2}R) > e^{\frac{1}{2}R\log R - c_7 R}. \tag{35}$$

Thus, for $R>4$, we obtain from (35) and (34),

$$\tfrac{1}{2}R\log R - c_7 R < \log M(R) < \tfrac{1}{2}R\log R + c_6 R,$$

which proves (31). Hence we obtain

THEOREM 7. *The function* ξ, *defined by* (11), *is an entire function of order* 1.

We can represent $\xi(s)$ as an infinite product by an appeal to a general result on entire functions of finite order. The proof of that result requires a special case of the following function-theoretical

LEMMA B (BOREL-CARATHÉODORY). *Suppose that the function* f *is regular in* $|z-z_0|<R$, *and has the expansion* $f(z)=\sum_{n=0}^{\infty} c_n(z-z_0)^n$. *Suppose further that* $\operatorname{Re} f(z)\leqslant U$ *for* $|z|<R$. *Then*

$$|c_n| \leqslant \frac{2(U-\operatorname{Re} c_0)}{R^n}, \qquad n=1,2,\ldots, \tag{36}$$

and, for $|z-z_0|\leqslant r<R$, *we have*

$$|f(z)-f(z_0)| \leqslant \frac{2r}{R-r}\Big\{U-\operatorname{Re} f(z_0)\Big\}, \tag{37}$$

as well as

$$\left|\frac{f^{(\nu)}(z)}{\nu!}\right| \leqslant \frac{2R}{(R-r)^{\nu+1}}\{U-\operatorname{Re} f(z_0)\}, \qquad \nu=1,2,3,\ldots. \tag{38}$$

PROOF. We may suppose, without loss of generality, that $z_0=0$ (cf. Lemma A). Let

$$\varphi(z)=U-f(z)=U-c_0-\sum_{n=1}^{\infty} c_n z^n=\sum_{n=0}^{\infty} b_n z^n, \tag{39}$$

say, for $|z|<R$. If $\mathscr{C}$ denotes the circle $|z|=r<R$, $r>0$, then

$$b_n = \frac{1}{2\pi i}\int_{\mathscr{C}} \frac{\varphi(z)\,dz}{z^{n+1}} = \frac{1}{2\pi r^n}\int_{-\pi}^{\pi} (P+iQ)e^{-in\theta}\,d\theta, \qquad n\geqslant 0, \tag{40}$$

where $\varphi(re^{i\theta})=P(r,\theta)+iQ(r,\theta)=P+iQ$, P, Q real. If we consider the integral of the regular function $\varphi(z)z^{n-1}, n\geqslant 1$, around $\mathscr{C}$, we obtain

$$0 = \frac{r^n}{2\pi}\int_{-\pi}^{\pi} (P+iQ)e^{in\theta}\,d\theta, \qquad n\geqslant 1.$$

Taking conjugates, and using (40), we get

$$b_n r^n = \frac{1}{\pi} \int_{-\pi}^{\pi} P e^{-in\theta} d\theta, \qquad n \geqslant 1.$$

Now $P = U - \operatorname{Re} f(z) \geqslant 0$ for $|z| < R$, and, in particular, on $\mathscr{C}$. Hence

$$|b_n| r^n \leqslant \frac{1}{\pi} \int_{-\pi}^{\pi} |P e^{-in\theta}| d\theta = \frac{1}{\pi} \int_{-\pi}^{\pi} P \, d\theta = 2 \operatorname{Re} b_0, \qquad n \geqslant 1,$$

by (40). Letting $r \to R$, we deduce that

$$|b_n| \leqslant \frac{2\beta_0}{R^n}, \qquad n \geqslant 1, \qquad \beta_0 = \operatorname{Re} b_0. \tag{41}$$

Now $b_n = -c_n$ for $n \geqslant 1$, and $b_0 = U - c_0$. Hence, for $|z| = r < R$, we have

$$|\varphi(z) - \varphi(0)| = \left| \sum_{n=1}^{\infty} b_n z^n \right| \leqslant \sum_{n=1}^{\infty} 2\beta_0 \left(\frac{r}{R} \right)^n = \frac{2\beta_0 r}{R - r},$$

by (41). Further

$$|\varphi^{(\nu)}(z)| \leqslant \sum_{n=\nu}^{\infty} n(n-1)(n-2)\dots(n-\nu+1) \cdot \frac{2\beta_0 r^{n-\nu}}{R^n}, \qquad \nu = 1, 2, \dots$$

$$= \left(\frac{d}{dr} \right)^{\nu} \left(\sum_{n=0}^{\infty} 2\beta_0 \left(\frac{r}{R} \right)^n \right) = \left(\frac{d}{dr} \right)^{\nu} \frac{2\beta_0 R}{R - r} = \frac{2\beta_0 \cdot R \cdot \nu!}{(R - r)^{\nu+1}},$$

which completes the proof of the lemma.

We shall use the special case $\nu = 1$ of (38) in the proof of the following

LEMMA C (HADAMARD). *Let $f(s)$ be an entire function of order ω, where ω is finite. Let $f(s)$ have an infinity of zeros $a_1, a_2, a_3, \dots$, (these being all the zeros) and $f(0) \neq 0$, so that*

$$0 < |a_1| \leqslant |a_2| \leqslant \dots |a_n| \to \infty,$$

zeros with the same modulus being arranged among themselves in any order, and multiple zeros being allowed for by repetition. Then

$$f(s) = e^{P(s)} \prod_{k=1}^{\infty} \left(1 - \frac{s}{a_k} \right) e^{\frac{s}{a_k} + \frac{s^2}{2a_k^2} + \dots + \frac{s^\mu}{\mu a_k^\mu}},$$

where $\mu = [\omega]$, and $P(s)$ is a polynomial in s of degree μ.

PROOF. Given $r>0$, let $n=n(r)$ be the integer defined by the inequalities $|a_n|\leqslant r$, $|a_{n+1}|>r$, so that $n(r)\to\infty$, as $r\to\infty$. Then $n(r)=O(r^{\omega+\varepsilon})$, for every $\varepsilon>0$, as $r\to\infty$. For we have only to take $z_0=0$, $R=2r_n$, $r=r_n$, with $r_n=|a_n|$, in Lemma A. Since the n zeros $a_1,a_2,\ldots,a_n$ lie in $|s|\leqslant r_n$, we have

$$2^n\leqslant \max_{|s|=2r_n}|f(s)|\cdot\frac{1}{|f(0)|},\qquad |f(0)|\neq 0,$$

or $n\log 2=O((2r_n)^{\omega+\varepsilon})$, which gives the stated result.

Now consider the integral

$$I=\frac{1}{2\pi i}\int\limits_{|z|=\frac{1}{2}R}\frac{f'(z)}{f(z)}\cdot\frac{1}{z^{\mu}(z-s)}\,dz,\qquad R>0.$$

We suppose that no zero a_k lies on the circle $|z|=\frac{1}{2}R$, and that $|s|<\frac{1}{8}R$, $s\neq a_k$ $(k=1,2,\ldots)$, $s\neq 0$.

By Cauchy's theorem, we have

$$I=\frac{f'(s)}{f(s)}\cdot\frac{1}{s^{\mu}}+\sum_{|a_k|<\frac{1}{2}R}\frac{1}{a_k^{\mu}(a_k-s)}+\frac{1}{(\mu-1)!}\left(\frac{d^{\mu-1}}{dz^{\mu-1}}\right)_{z=0}\left(\frac{f'(z)}{f(z)\,(z-s)}\right).$$

We shall prove that the integral I on the left-hand side tends to zero as $R\to\infty$ through such a sequence of values as will ensure that the corresponding circles $|z|=\frac{1}{2}R$ are all free from any zero of f. For that purpose we write

$$2\pi i I=\int\limits_{|z|=\frac{1}{2}R}\left(\frac{f'(z)}{f(z)}-\sum_{k=1}^{N}\frac{1}{z-a_k}\right)\frac{dz}{z^{\mu}(z-s)}+\sum_{k=1}^{N}\int\limits_{|z|=\frac{1}{2}R}\frac{1}{z-a_k}\cdot\frac{dz}{z^{\mu}(z-s)}$$

$$=I_1+I_2,$$

say, where N is the largest integer such that $|a_N|\leqslant R$. If M is the largest integer such that $|a_M|<\frac{1}{2}R$ (there being no a_k such that $|a_k|=\frac{1}{2}R$ by assumption), we have

$$I_2=\left(\sum_{k=1}^{M}+\sum_{k=M+1}^{N}\right)\int\limits_{|z|=\frac{1}{2}R}\frac{1}{z-a_k}\cdot\frac{dz}{z^{\mu}(z-s)}$$

$$=J_1+J_2,$$

say. The value of J_1 is unchanged, if we integrate along the circle $|z|=\frac{3}{4}R$ instead of $|z|=\frac{1}{2}R$, since there are no poles of the integrand in the annulus $\frac{1}{2}R\leqslant|z|\leqslant\frac{3}{4}R$. Along the new path, we have $|z-a_k|\geqslant\frac{1}{4}R$ for $1\leqslant k\leqslant M$, and since $N=O(R^{\omega+\varepsilon})$, $\varepsilon>0$, we have

$$J_1=O(R^{\omega-\mu-1+\varepsilon})=o(1),$$

as $R\to\infty$, if $0<\varepsilon<\mu+1-\omega$. Similarly the value of J_2 is unchanged if the path of integration is changed to the circle $|z|=\frac{1}{4}R$, and we get

$$J_2=o(1),$$

as $R\to\infty$.

We next prove that $I_1=o(1)$, as $R\to\infty$, by establishing the inequality

$$\left|\frac{f'(z)}{f(z)}-\sum_{k=1}^{N}\frac{1}{z-a_k}\right|<cR^{\omega-1+\varepsilon}, \qquad 0<\varepsilon<1, \tag{42}$$

where c is a constant, and $|z|=\frac{1}{2}R$.

Let

$$g_R(z)=\frac{f(z)}{f(0)}\prod_{k=1}^{N}\left(1-\frac{z}{a_k}\right)^{-1},$$

where $f(0)\neq 0$, by assumption. Then $g_R(z)$ is regular, and $g_R(z)\neq 0$, for $|z|\leqslant R$. Further $g_R(0)=1$, so that we can write

$$g_R(z)=e^{h_R(z)},$$

where $h_R(z)$ is regular for $|z|\leqslant R$, $h_R(0)=0$.

If $|z|=2R$, we have, for every $\varepsilon>0$,

$$|f(z)|=O(e^{(2R)^{\omega+\varepsilon}}), \qquad \left|1-\frac{z}{a_k}\right|\geqslant 1, \quad \text{for} \quad 1\leqslant k\leqslant N,$$

so that

$$|g_R(z)|=O(e^{(2R)^{\omega+\varepsilon}}), \quad |z|=2R, \qquad 0<\varepsilon<1.$$

By the maximum modulus principle, this holds also for $|z|=R$. Hence

$$\operatorname{Re} h_R(z)=\log|g_R(z)|=O(R^{\omega+\varepsilon}), \qquad |z|=R.$$

By (38) with $\nu=1$, we have, for $|z|=\frac{1}{2}R$,

$$|h_R'(z)|=O(R^{\omega-1+\varepsilon}),$$

which gives (42). It is immediate that

$$I_1=O(R^{\omega-\mu-1+\varepsilon})=o(1),$$

as $R\to\infty$, if $0<\varepsilon<\mu+1-\omega$.

Thus

$$\frac{f'(s)}{f(s)}\cdot\frac{1}{s^{\mu}}+\frac{1}{(\mu-1)!}\left(\frac{d^{\mu-1}}{dz^{\mu-1}}\right)_{z=0}\left(\frac{f'(z)}{f(z)}\cdot\frac{1}{z-s}\right)+\sum_{k=1}^{\infty}\frac{1}{a_k^{\mu}(a_k-s)}=0.$$

But

$$\frac{1}{(\mu-1)!}\left(\frac{d^{\mu-1}}{dz^{\mu-1}}\right)_{z=0}\left(\frac{f'(z)}{f(z)}\cdot\frac{1}{z-s}\right)=\sum_{k=1}^{\mu}c_k s^{-k},$$

where

$$c_k = -\frac{\varphi^{(\mu-k)}(0)}{(\mu-k)!}, \qquad \varphi(z) = \frac{f'(z)}{f(z)}.$$

Hence

$$\frac{f'(s)}{f(s)} = \sum_{k=1}^{\infty} \left(\frac{s}{a_k}\right)^{\mu} \frac{1}{s-a_k} + \sum_{k=0}^{\mu-1} c'_k s^k, \qquad c'_k = -c_{\mu-k}. \tag{43}$$

Since

$$\left(\frac{s}{a_k}\right)^{\mu} \frac{1}{s-a_k} = \left(\frac{1}{s} + \frac{a_k}{s^2} + \cdots + \frac{a_k^{\mu-1}}{s^{\mu}} + \frac{1}{(s-a_k)}\left(\frac{a_k}{s}\right)^{\mu}\right)\left(\frac{s}{a_k}\right)^{\mu},$$

we obtain, by integrating (43) and raising to the power of e,

$$f(s) = e^{\sum_{k=0}^{\mu} d_k s^k} \prod_{k=1}^{\infty} \left(1 - \frac{s}{a_k}\right) e^{\frac{s}{a_k} + \frac{s^2}{2a_k^2} + \cdots + \frac{s^{\mu}}{\mu a_k^{\mu}}}, \qquad f(0) = e^{d_0}.$$

This holds also for $s = a_k$ and for $s = 0$.

REMARK. If $f(0) = 0$, then we write $f(s) = s^{\alpha}\psi(s)$, $\psi(0) \neq 0$, and apply the lemma to ψ.

By an application of Lemma C, we obtain from Theorem 7, the following representation:

$$\xi(s) = e^{b_0 + b_1 s} \prod_{\rho} \left(1 - \frac{s}{\rho}\right) e^{s/\rho}, \tag{44}$$

where ρ runs through the zeros of ξ, and b_0, b_1 are constants. By logarithmic differentiation, we obtain

$$\frac{\xi'(s)}{\xi(s)} = b_1 + \sum_{\rho} \left(\frac{1}{s-\rho} + \frac{1}{\rho}\right), \tag{45}$$

the dash denoting the derivative. This leads to the formula

$$\frac{\zeta'(s)}{\zeta(s)} = b_1 + \frac{1}{2}\log\pi - \frac{1}{s-1} - \frac{1}{2}\frac{\Gamma'}{\Gamma}\left(\frac{1}{2}s+1\right) + \sum_{\rho}\left(\frac{1}{s-\rho} + \frac{1}{\rho}\right), \tag{46}$$

in view of (10) and (11).

§ 4. Hardy's theorem. It follows from Theorem 6 that the function ζ has an infinity of non-real zeros in the critical strip. We shall prove that an infinity of them lie on the critical line. The proof requires the estimation of some exponential integrals.

Let $C^k[a,b]$ denote the class of real functions in the interval $a \leqslant x \leqslant b$, which are k times continuously differentiable.

Lemma D. *Let* $F \in C^1[a,b]$, *and let* F', *the first derivative of* F, *be monotone, and* $|F'(x)| \geqslant m > 0$, *throughout the interval* $a \leqslant x \leqslant b$. *Then*

$$\left| \int_a^b e^{iF(x)}\,dx \right| \leqslant \frac{4}{m}.$$

Proof. We can assume, by taking conjugates if necessary, that F' is positive. Taking the real and imaginary parts of the integral separately, we see that

$$\int_a^b \cos(F(x))\,dx = \int_a^b \frac{F'(x)\cos(F(x))}{F'(x)}\,dx,$$

and an application of the *second mean-value theorem* gives

$$\left| \int_a^b \cos(F(x))\,dx \right| \leqslant \frac{2}{m}.$$

Similarly we have

$$\left| \int_a^b \sin(F(x))\,dx \right| \leqslant \frac{2}{m},$$

and hence the lemma.

Lemma E. *Let* $F, G \in C^2[a,b]$, $G \neq 0$, *and* $(F'/G)'$ *have at most* q *distinct zeros in the interval* $a \leqslant x \leqslant b$. *Let*

$$\left| \frac{F'(x)}{G(x)} \right| \geqslant m > 0,$$

throughout $[a,b]$. *Then*

$$\left| \int_a^b G(x) e^{iF(x)}\,dx \right| \leqslant \frac{4(q+1)}{m}.$$

Proof. We divide the interval $[a,b]$ into (at most) $q+1$ intervals in each of which G/F' is monotone, and apply an argument similar to that of Lemma D to each of them.

Lemma F. *Let* $F \in C^2[a,b]$, *and* $|F''(x)| \geqslant r > 0$, *throughout* $[a,b]$. *Then*

$$\left| \int_a^b e^{iF(x)}\,dx \right| \leqslant \frac{8}{\sqrt{r}}.$$

Proof. We may assume, as before, that $F''(x) \geqslant r > 0$, which implies that F' is monotone increasing, and therefore vanishes at most once in the interval $[a,b]$, say at c. Let $\delta>0$, and denote by I, I_1, I_2, I_3, the following intervals:

$$I=[a,b]\,; \qquad I_1=\begin{cases}[a,c-\delta], & \text{if } c-\delta>a,\\ \emptyset, & \text{if } c-\delta\leqslant a;\end{cases}$$

$$I_2=[c-\delta,c+\delta]\cap I\,; \quad I_3=\begin{cases}[c+\delta,b], & \text{if } c+\delta<b,\\ \emptyset, & \text{if } c+\delta\geqslant b.\end{cases}$$

We then have

$$\int_a^b e^{iF(x)}dx=\int_I e^{iF(x)}dx=\int_{I_1}+\int_{I_2}+\int_{I_3}.$$

It is immediate that

$$\left|\int_{I_2}\right|\leqslant 2\delta,$$

while in I_3, we have, if it is not empty,

$$F'(x)=\int_c^x F''(t)\,dt\geqslant r(x-c)\geqslant r\delta,$$

so that Lemma D gives

$$\left|\int_{I_3}\right|\leqslant\frac{4}{r\delta},$$

and similarly, if I_1 is not empty,

$$\left|\int_{I_1}\right|\leqslant\frac{4}{r\delta}.$$

Hence

$$\left|\int_I\right|\leqslant\frac{8}{r\delta}+2\delta,$$

and if we choose $\delta=\dfrac{2}{\sqrt{r}}$, we get Lemma F. If F' does not vanish in $[a,b]$, the argument is similar, with c replaced by a or b according as F' is positive or negative in $[a,b]$.

LEMMA G. *Let $F\in C^2[a,b]$, and $|F''(x)|\geqslant r>0$, throughout $[a,b]$. Let $G\in C^2[a,b]$; $|G(x)|\leqslant M$, for $a\leqslant x\leqslant b$; $G\neq 0$; and $(F'/G)'$ have at most q distinct zeros in $[a,b]$. Then*

$$\left|\int_a^b G(x)e^{iF(x)}dx\right|\leqslant\frac{8M(q+1)}{\sqrt{r}}. \tag{47}$$

PROOF. The argument runs along the same lines as in Lemma F, except that we now use Lemma E instead of Lemma D.

We are now in a position to prove

THEOREM 8 (HARDY). *There exist an infinity of non-real zeros of* ζ *with real part* $\frac{1}{2}$.

PROOF (HARDY-LITTLEWOOD). The functional equation of ζ, given by (3), can be written in the form $\zeta(s)=\chi(s)\zeta(1-s)$, where

$$\chi(s) = \frac{\pi^{s-\frac{1}{2}}\Gamma(\frac{1}{2}-\frac{1}{2}s)}{\Gamma(\frac{1}{2}s)}.$$

It follows that $\chi(s)\chi(1-s)=1$, which implies that $\chi(\frac{1}{2}+it)\chi(\frac{1}{2}-it)=1$, and from this we can infer that $|\chi(\frac{1}{2}+it)|=1$, since $\chi(s)$ is real for real s. Let

$$\theta=\theta(t)=-\tfrac{1}{2}\arg\chi(\tfrac{1}{2}+it),$$

so that

$$\chi(\tfrac{1}{2}+it)=e^{-2i\theta}.$$

Define

$$Z(t)=e^{i\theta}\zeta(\tfrac{1}{2}+it)=(\chi(\tfrac{1}{2}+it))^{-\frac{1}{2}}\zeta(\tfrac{1}{2}+it). \tag{48}$$

Since $\Gamma(s)$ has no zeros and only real poles, the function $(\chi(s))^{-1}$ has a square-root $(\chi(s))^{-\frac{1}{2}}$ in the simply connected region $t>0$. We can rewrite (48) as follows:

$$\begin{aligned} Z(t)&=\pi^{-\frac{1}{2}it}\left\{\frac{\Gamma(\frac{1}{4}+\frac{1}{2}it)}{\Gamma(\frac{1}{4}-\frac{1}{2}it)}\right\}^{\frac{1}{2}}\zeta(\tfrac{1}{2}+it)\\ &=\pm\pi^{-\frac{1}{2}it}\frac{\Gamma(\frac{1}{4}+\frac{1}{2}it)}{|\Gamma(\frac{1}{4}+\frac{1}{2}it)|}\zeta(\tfrac{1}{2}+it). \end{aligned} \tag{49}$$

We choose the $+$ sign, and keep it fixed in what follows. If $s=\frac{1}{2}+iz$, where z is complex, and

$$\Xi(z)=\xi(\tfrac{1}{2}+iz), \tag{50}$$

then

$$\Xi(z)=\xi(\tfrac{1}{2}+iz)=\xi(\tfrac{1}{2}-iz)=\Xi(-z).$$

By Theorem 3, $\xi(\frac{1}{2}+iz)$ is real for real z, so that $\Xi(t)$ is real for real t. Since

$$\begin{aligned} \Xi(t)&=\tfrac{1}{2}(\tfrac{1}{2}+it)(-\tfrac{1}{2}+it)\eta(\tfrac{1}{2}+it)\\ &=-\frac{(t^2+\frac{1}{4})\Gamma(\frac{1}{4}+\frac{1}{2}it)\zeta(\frac{1}{2}+it)}{2\pi^{\frac{1}{4}+\frac{1}{2}it}}, \end{aligned}$$

it follows from (49) that

$$Z(t) = -\frac{2\pi^{\frac{1}{4}}\Xi(t)}{(t^2+\frac{1}{4})\,|\Gamma(\frac{1}{4}+\frac{1}{2}it)|},$$

and that

$$Z(t) \quad \text{is real for real } t, \tag{51}$$

and, because of (48), that

$$|Z(t)| = |\zeta(\tfrac{1}{2}+it)|. \tag{52}$$

If $\zeta(s)$ had no zeros, or only a finite number of zeros, on the line $\sigma = \frac{1}{2}$, then there would exist a T_0, such that for $T > T_0$, we have

$$\left|\int_T^{2T} Z(t)\,dt\right| = \int_T^{2T} |Z(t)|\,dt.$$

We shall prove, in fact, that that is not the case.

Consider the integral

$$\int_{\mathscr{C}} \left\{\chi(s)\right\}^{-\frac{1}{2}} \zeta(s)\,ds, \tag{53}$$

taken along the rectangle $\mathscr{C}$, bounded by the lines $\sigma = \frac{1}{2}$, $\sigma = \frac{5}{4}$, $t = T$, and $t = 2T$, where $T > c > 0$. The integral vanishes, by Cauchy's theorem. The contribution from that part of $\mathscr{C}$ which lies on $\sigma = \frac{1}{2}$ is

$$-\int_{\frac{1}{2}+iT}^{\frac{1}{2}+2iT} \left\{\chi(s)\right\}^{-\frac{1}{2}} \zeta(s)\,ds = -i\int_T^{2T} Z(t)\,dt, \tag{54}$$

because of the definition of $Z(t)$ given in (48). To estimate the contributions from the other three sides of the rectangle, we use Stirling's formula for the gamma-function in the following form (cf. (23)): in any fixed strip $-\infty < a \leqslant \sigma \leqslant b < +\infty$, we have, as $t \to \infty$,

$$\Gamma(\sigma+it) = t^{\sigma+it-\frac{1}{2}} e^{-\frac{1}{2}\pi t - it + \frac{1}{2}i\pi(\sigma-\frac{1}{2})} (2\pi)^{\frac{1}{2}} \left\{1 + O\left(\frac{1}{t}\right)\right\}. \tag{55}$$

We then obtain

$$\begin{aligned}\left\{\chi(s)\right\}^{-\frac{1}{2}} &= \pm\left(\frac{1}{\pi}\right)^{\frac{1}{2}\sigma-\frac{1}{4}+\frac{1}{2}it} \left(\frac{\Gamma(\frac{1}{2}s)}{\Gamma(\frac{1}{2}-\frac{1}{2}s)}\right)^{\frac{1}{2}} \\ &= \pm\left(\frac{t}{2\pi}\right)^{\frac{1}{2}\sigma-\frac{1}{4}+\frac{1}{2}it} e^{-\frac{1}{2}it-\frac{1}{8}i\pi} \left\{1 + O\left(\frac{1}{t}\right)\right\}.\end{aligned} \tag{56}$$

On the other hand, we have, by (14),

$$\zeta(s)=O(\log t)=O(t^{\varepsilon}), \qquad \varepsilon>0,$$

for $\sigma=1$, $t\geqslant 2$, so that

$$\overline{\zeta(it)}=\zeta(-it)=\frac{\zeta(1+it)}{\chi(1+it)}=O(t^{\frac{1}{2}}\log t).$$

Hence, by the Phragmén-Lindelöf principle, we have

$$\zeta(s)=O\left(t^{\frac{1}{2}-\frac{1}{2}\sigma+\varepsilon}\right), \qquad \tfrac{1}{2}\leqslant\sigma\leqslant 1.$$

Similarly the fact that $\zeta(s)=O(1)$ for $\sigma=\frac{5}{4}$, and $\zeta(s)=O(t^{\varepsilon})$ for $\sigma=1$, imply that

$$\zeta(s)=O(t^{\varepsilon}), \qquad 1<\sigma\leqslant\tfrac{5}{4}.$$

(This follows also from (14)).

Hence

$$\left\{\chi(s)\right\}^{-\frac{1}{2}}\zeta(s)=\begin{cases}O\left(t^{\frac{1}{2}\sigma-\frac{1}{4}}\cdot t^{\frac{1}{2}-\frac{1}{2}\sigma+\varepsilon}\right)=O\left(t^{\frac{1}{4}+\varepsilon}\right), & \tfrac{1}{2}\leqslant\sigma\leqslant 1,\\ O\left(t^{\frac{1}{2}\sigma-\frac{1}{4}+\varepsilon}\right)=O\left(t^{\frac{3}{8}+\varepsilon}\right), & 1<\sigma\leqslant\tfrac{5}{4}.\end{cases}$$

Thus the contributions to the integral (53) from the sides of the rectangle $\mathscr{C}$ parallel to the real axis are $O(T^{\frac{3}{8}+\varepsilon})$, $\varepsilon>0$.

The integral along the line $\sigma=\frac{5}{4}$ gives, because of (56),

$$\pm i\int_{T}^{2T}\left(\frac{t}{2\pi}\right)^{\frac{3}{8}+\frac{1}{2}it} e^{-\frac{1}{2}it-\frac{1}{8}i\pi}\left\{1+O\left(\frac{1}{t}\right)\right\}\zeta(\tfrac{5}{4}+it)\,dt.$$

The contribution of the O-term is

$$\int_{T}^{2T}O(t^{-\frac{5}{8}})\,dt=O(T^{\frac{3}{8}}),$$

while the other term is a constant multiple of

$$\sum_{n=1}^{\infty}n^{-\frac{5}{4}}\int_{T}^{2T}\left(\frac{t}{2\pi}\right)^{\frac{3}{8}+\frac{1}{2}it}e^{-\frac{1}{2}it-it\log n}\,dt.$$

Since

$$\frac{d^2}{dt^2}\left(\frac{1}{2}t\log\frac{t}{2\pi}-\frac{1}{2}t-t\log n\right)=\frac{1}{2t},$$

we can apply Lemma G, and prove that the integral in the above series is $O(T^{\frac{7}{8}})$, uniformly in n, so that the sum of the series is $O(T^{\frac{7}{8}})$.

Combining all the estimates, we have

$$\left|\int_T^{2T} Z(t)\,dt\right| = O(T^{\frac{7}{8}}). \tag{57}$$

On the other hand, we have

$$\int_T^{2T} |Z(t)|\,dt = \int_T^{2T} |\zeta(\tfrac{1}{2}+it)|\,dt \geqslant \left|\int_T^{2T} \zeta(\tfrac{1}{2}+it)\,dt\right|,$$

and

$$i\int_T^{2T} \zeta(\tfrac{1}{2}+it)\,dt = \int_{\frac{1}{2}+iT}^{\frac{1}{2}+2iT} \zeta(s)\,ds = \int_{\frac{1}{2}+iT}^{2+iT} + \int_{2+iT}^{2+2iT} + \int_{2+2iT}^{\frac{1}{2}+2iT}$$

$$= \left[s - \sum_{n=2}^{\infty} \frac{1}{n^s \log n}\right]_{2+iT}^{2+2iT} + \int_{\frac{1}{2}}^{2} O(T^{\frac{1}{2}})\,d\sigma = iT + O(T^{\frac{1}{2}}).$$

Hence there exists a T_0, such that

$$\int_T^{2T} |Z(t)|\,dt > \tfrac{1}{2}T, \quad \text{for} \quad T > T_0. \tag{58}$$

From (57) and (58) it follows that $Z(t)$ cannot ultimately be of the same sign, and hence the theorem.

REMARKS. It was conjectured by Riemann that all the non-real zeros of ζ lie on the critical line. This is referred to as the *Riemann hypothesis*, and is yet to be proved, or disproved.

If $N_0(T)$ denotes the number of zeros of ζ of the form $\frac{1}{2}+it$, $0<t\leqslant T$, then Hardy and Littlewood proved that $N_0(T)>cT$, for a positive constant c, and for large T, in comparison with the result $N(T)\sim\dfrac{1}{2\pi}T\log T$ proved in Theorem 6. The best result so far known is the one obtained by Atle Selberg, namely $N_0(T)>cT\log T$.

§ 5. Hamburger's theorem. The functional equation characterizes the zeta-function of Riemann to the extent described by the following

THEOREM 9 (HAMBURGER). *Let G be an entire function of finite order, and P a polynomial, and s a complex variable, written $s=\sigma+it$, σ and t real, $i^2=-1$. Let $f(s)=G(s)/P(s)$, and*

$$f(s) = \sum_{n=1}^{\infty} \frac{a_n}{n^s},$$

where (a_n) *is a sequence of complex numbers, and the series converges absolutely for* $\sigma>1$. *Let*

$$\pi^{-\frac{1}{2}s}\Gamma(\tfrac{1}{2}s)\,f(s)=\pi^{-\frac{1}{2}(1-s)}\Gamma(\tfrac{1}{2}-\tfrac{1}{2}s)g(1-s), \tag{59}$$

where

$$g(1-s)=\sum_{n=1}^{\infty}\frac{b_n}{n^{1-s}},$$

the series converging absolutely for $\sigma<-\alpha<0$. *Then* $f(s)=a_1\zeta(s)=g(s)$.

PROOF (SIEGEL). For $x>0$, let

$$S_1=\frac{1}{2\pi i}\int_{2-i\infty}^{2+i\infty} f(s)\Gamma(\tfrac{1}{2}s)\pi^{-s/2}x^{-s/2}\,ds, \tag{60}$$

the integral converging since $f(s)$ is bounded on the line $\sigma=2$. Since the series $\sum_{n=1}^{\infty}a_n n^{-s}$ converges absolutely for $\sigma>1$, we have

$$S_1=\sum_{n=1}^{\infty}\frac{a_n}{2\pi i}\int_{2-i\infty}^{2+i\infty}\Gamma(\tfrac{1}{2}s)(\pi n^2 x)^{-s/2}\,ds=2\sum_{n=1}^{\infty}a_n e^{-\pi n^2 x}, \tag{61}$$

if we note that

$$e^{-y}=\frac{1}{2\pi i}\int_{1-i\infty}^{1+i\infty}y^{-s}\Gamma(s)\,ds, \qquad y>0. \tag{62}$$

By (59) we have $S_1=S_2$, where

$$S_2=\frac{1}{2\pi i}\int_{2-i\infty}^{2+i\infty}g(1-s)\Gamma(\tfrac{1}{2}-\tfrac{1}{2}s)\pi^{-\frac{1}{2}(1-s)}x^{-s/2}\,ds. \tag{63}$$

We shall move the line of integration in (63) from $\sigma=2$ to the line $\sigma=-1-\alpha<-1$, where α is such that all the poles of f (which are finite in number) are contained in the strip $-1-\alpha<\sigma<2$. The function $g(1-s)$ is bounded on the line $\sigma=-1-\alpha$, since the series $\sum_{n=1}^{\infty}b_n n^{+s-1}$ converges absolutely for $\sigma<-\alpha$. Since $f(s)$ is bounded on the line $\sigma=2$, and

$$\frac{\Gamma(\tfrac{1}{2}s)}{\Gamma(\tfrac{1}{2}-\tfrac{1}{2}s)}=O(|t|^{\sigma-\frac{1}{2}}),$$

it follows from (59) that $g(1-s)=O(|t|^{\frac{3}{2}})$ on $\sigma=2$. By the hypothesis on f, it follows that there exist two numbers $T>0$, $\rho>0$, such that for $|t|\geqslant T$ and $-1-\alpha\leqslant\sigma\leqslant 2$, the function $g(1-s)$ is regular and $g(1-s)=O(e^{|t|^{\rho}})$. Hence, by the Phragmén-Lindelöf principle,

$$g(1-s)=O\,(|t|^{\frac{3}{2}}),$$

for $|t|\geqslant T$, $-\alpha-1\leqslant\sigma\leqslant 2$. By applying Cauchy's theorem to a suitably chosen rectangle, we therefore obtain

$$S_2=\frac{1}{2\pi i}\int_{-1-\alpha-i\infty}^{-1-\alpha+i\infty} g(1-s)\Gamma(\tfrac{1}{2}-\tfrac{1}{2}s)\pi^{-\frac{1}{2}(1-s)}x^{-s/2}\,ds+\sum_{\nu=1}^{m}R_\nu, \tag{64}$$

where $\sum_{\nu=1}^{m}R_\nu$ denotes the sum of the residues of the integrand at all of its poles. But the integrand is $f(s)\Gamma(\frac{1}{2}s)\pi^{-\frac{1}{2}s}x^{-s/2}$, and its residue at a pole s_ν of order q_ν is of the form

$$x^{-s_\nu/2}(A_{q_\nu-1}^{(\nu)}\log^{q_\nu-1}x+\cdots+A_1^{(\nu)}\log x+A_0^{(\nu)}).$$

Hence

$$\sum_{\nu=1}^{m}R_\nu=\sum_{\nu=1}^{m}x^{-s_\nu/2}Q_\nu(\log x)=Q(x), \tag{65}$$

say, where $Q_\nu(x)$ is a polynomial in x. Here

$$\operatorname{Re}s_\nu\leqslant 2-\delta,\qquad \delta>0,\quad \text{for}\quad \nu=1,2,\ldots,m, \tag{66}$$

since $\sum_{n=1}^{\infty}a_n n^{-s}$ converges absolutely for $\sigma>2-\varepsilon$, for a sufficiently small $\varepsilon>0$. Hence we have, by (64),

$$\begin{aligned}S_2&=\frac{1}{\sqrt{x}}\sum_{n=1}^{\infty}\frac{b_n}{2\pi i}\int_{-1-\alpha-i\infty}^{-1-\alpha+i\infty}\Gamma(\tfrac{1}{2}-\tfrac{1}{2}s)\left(\frac{\pi n^2}{x}\right)^{-\frac{1}{2}(1-s)}ds+Q(x)\\&=\frac{2}{\sqrt{x}}\sum_{n=1}^{\infty}\frac{b_n}{2\pi i}\int_{1+\frac{1}{2}\alpha-i\infty}^{1+\frac{1}{2}\alpha+i\infty}\Gamma(s)\left(\frac{\pi n^2}{x}\right)^{-s}ds+Q(x)\\&=\frac{2}{\sqrt{x}}\sum_{n=1}^{\infty}b_n e^{-\pi n^2/x}+Q(x),\end{aligned} \tag{67}$$

by (62). Since $S_1=S_2$, we have, from (61) and (67),

$$2\sum_{n=1}^{\infty}a_n e^{-\pi n^2 x}=\frac{2}{\sqrt{x}}\sum_{n=1}^{\infty}b_n e^{-\pi n^2/x}+Q(x). \tag{68}$$

If we multiply throughout by $e^{-\pi t^2 x}$, where t is fixed, $t>0$, and integrate with respect to x over $(0,\infty)$, we have

$$2\sum_{n=1}^{\infty}\frac{a_n}{\pi(t^2+n^2)}=2\sum_{n=1}^{\infty}\frac{b_n}{t}e^{-2\pi nt}+\int_0^{\infty}Q(x)e^{-\pi t^2 x}\,dx, \tag{69}$$

since

$$\int_0^{\infty}e^{-a^2x-b^2/x}x^{-\frac{1}{2}}\,dx=\frac{\pi^{\frac{1}{2}}}{a}e^{-2ab}, \quad \text{for} \quad a>0, \quad b\geqslant 0. \tag{70}$$

We can substitute for $Q(x)$ in (69) from (65), and integrate term by term, since each of the terms is $O(x^{-1+\frac{1}{4}\delta})$ as $x\to 0$, because of (66). Hence

$$\int_0^{\infty}Q(x)e^{-\pi t^2 x}\,dx=\frac{1}{t^2}\int_0^{\infty}Q\left(\frac{x}{t^2}\right)e^{-\pi x}\,dx=\sum_{\nu=1}^{m}t^{s_\nu-2}H_\nu(\log t)=H(t), \tag{71}$$

say, where $H_\nu(\log t)$ is a polynomial in $\log t$. By (69) and (71) we have, for $t>0$,

$$\sum_{n=1}^{\infty}a_n\left(\frac{1}{t+in}+\frac{1}{t-in}\right)-\pi t H(t)=2\pi\sum_{n=1}^{\infty}b_n e^{-2\pi nt}. \tag{72}$$

If we now consider the complex t-plane, the series on the left converges uniformly in every compact set excluding the points $t=\pm ik$, $k=1,2,\ldots$. It therefore represents a meromorphic function with simple poles at $t=\pm ik$, with residue a_k, if $a_k\neq 0$, otherwise regular at $t=\pm ik$. The function $H(t)$ is (single-valued and) regular in the t-plane with the negative real axis $(-\infty,0]$ deleted. The right-hand side of (72) is regular for $\operatorname{Re} t>0$, since the series $\sum_{n=1}^{\infty}b_n n^{-s}$ converges absolutely for $\sigma>1+\alpha$, and is a periodic function of t with period i. Hence the residues at the points ki and $(k+1)i$ are equal. Therefore $a_k=a_{k+1}$, $k=1,2,\ldots$, which implies that $f(s)=a_1\zeta(s)$.

Notes on Chapter II

As general references, see E. Landau's *Vorlesungen*, II, 63–95; E. C. Titchmarsh's *Zeta-function*, 13–37; 178–181; 219–220; and Ingham's *Tract*, Ch. III.

§ 1. For Euler's identity, see for example, the author's *Introduction*, Ch. VII, § 4. For elementary facts about Dirichlet series, see ibid. Ch. X, § 4.

Bochner's proof of the Poisson summation formula is given in his book, *Vorlesungen über Fouriersche Integrale*, (Leipzig), 1932, 37. Summability of an infinite series by arithmetical means is referred to as summability (C, 1), that is by Cesàro's means of the first order. For a proof of Fejér's theorem for Fourier series, see, for instance, E. C. Titchmarsh's *Theory of functions*, 2nd edition (1939), 414. The convergence of a series trivially implies summability (C, 1). Formula (6) can, of course, be proved directly. See Landau's *Vorlesungen*, II, Satz 415.

There are many proofs of the functional equation of the zeta-function of Riemann. Some are given, for instance, in Titchmarsh's *Zeta-function*, 13–37. The proof given here is one of Riemann's original proofs. See his paper, Über die Anzahl der Primzahlen unter einer gegebenen Grösse, *Monatsberichte der Preuss. Akad. der Wissenschaften* (Berlin, 1859–60), 671–680; *Werke* (1st edition, 1876), 136–144; (2nd edition, 1892), 145–155; *Oeuvres* (1898), 165–176.

If $f(x)$ is a Lebesgue integrable function in $(-\infty, \infty)$, its Fourier transform is defined as

$$\varphi(\alpha) = \int_{-\infty}^{\infty} f(x)\, e^{i\alpha x}\, dx$$

for real α. The theta-relation (6) follows from Theorem 1 if we use the fact that $f(x) = e^{-x^2}$ implies that $\varphi(\alpha) = \pi^{\frac{1}{2}} e^{-\alpha^2/4}$. This in turn follows from the fact that $\varphi'(\alpha) = -\frac{1}{2}\alpha\varphi(\alpha)$, while $\varphi(0) = \pi^{\frac{1}{2}}$.

Alternatively one can prove the theta-relation directly by contour integration. See E. Landau's *Primzahlen*, § 69, or his *Vorlesungen*, II, Satz 415.

For the proof that $\dfrac{1}{\Gamma(s)}$ is an entire function, and for other properties of the gamma-function, see Titchmarsh's *Theory of functions*, loc. cit., §§ 4.41, 4.42, § 8.4.

Another proof of the functional equation conceived by Riemann is fully developed by Siegel, *Quellen und Studien zur Geschichte der Math. Astr. und Physik*, Abt. B: Studien 2 (1932), 45–80; *Gesammelte Abhandlungen* (1966), I, 275–310.

The proof of Theorem 3 uses the principle of reflection. See, for instance, L. V. Ahlfors's *Complex analysis*, 2nd edition, McGraw-Hill (1966), 170. If ξ is entire, real on the real axis, then $\xi(s)$ and $\overline{\xi(\bar{s})}$ are both entire, coincide for real s, so coincide everywhere.

The treatment of Theorems 3 and 4 follows Ingham's exposition in his *Tract*, Ch. III, §§ 3, 4, 5.

§ 2. For Abel's summation formula, see the Notes on Chapter I.

Lemma A, given in Ingham's *Tract*, Ch. III, § 6, is a consequence of Jensen's formula. See, for instance, Ahlfors's *Complex analysis*, loc. cit. 205. For the maximum modulus principle, see ibid. 133.

Theorem 6 was stated by Riemann (loc. cit.), but first proved by H. von Mangoldt, *Math. Annalen*, 60 (1905), 1 – 19. See Landau's *Primzahlen*, I, §§ 90 – 92. Backlund's proof given here is in *Acta Math.* 41 (1918), 345 – 375; see Ingham's *Tract*, Ch. IV, § 2. The "variation of the argument" is avoided here, as a result of a suggestion made by Prof. Raghavan Narasimhan.

§ 3. For elementary properties of entire (= integral) functions, see Titchmarsh's *Theory of functions*, loc. cit., Ch. VIII.

For Lemma B see Landau's *Vorlesungen*, I, Satz 225; and Ingham's *Tract*, 50. Inequality (36) is due to E. Borel, and (38) to C. Carathéodory. See J. E. Littlewood's *Theory of functions* (Oxford, 1944), 114 – 116.

Lemma C is due to J. Hadamard, *Journal de Math.* (4), 9 (1893), 171 – 215. Landau gave a proof of it in *Math. Zeitschrift*, 26 (1927), 170 – 175. The proof given here, which does not assume any knowledge of the Weierstrass product, is due to the author, *J. Indian Math. Soc.* 5 (1941), 128 – 132. The same idea can be extended to R. Nevanlinna's factorization theorem for meromorphic functions (where the numerator and denominator do not necessarily converge separately), see *Journal of the Madras University*, XV (1943), Sec. B, 11 – 17; R. Nevanlinna's *Eindeutige analytische Funktionen* (Berlin, 1936), 213.

§ 4. Lemmas D, E, F, G have long been known. See, for instance, Landau's *Vorlesungen*, II, Satz 413, and his remarks on Satz 429. They are treated here as in Titchmarsh's *Zeta-function*, §§ 4.2 – 4.5. For the 'second mean-value theorem' of the integral calculus, see for instance G. H. Hardy's *A course of pure mathematics*, 10th edition (Cambridge, 1952), 325 – 326.

Theorem 8 is due to G. H. Hardy, *Comptes Rendus*, Paris, 158 (1914), 1012 – 14. The proof given here is due to G. H. Hardy and J. E. Littlewood, and is published in Landau's *Vorlesungen*, II, 78 – 85. Landau exclaims: "Ist dieser (Hardy-Littlewoodsche) Beweis nicht schön?" Other proofs are also given in Titchmarsh's *Zeta-function*, §§ 10.2 – 10.5. This proof extends to the zeta-function of an ideal class in a real or imaginary quadratic field. See, for instance, the paper by the author and Raghavan Narasimhan, *Commentarii Mathematici Helvetici*, 43 (1968), 18 – 30.

The proof that $N_0(T) > cT$ requires a slightly more detailed analysis of the proof of Theorem 8, but no new principle. See Titchmarsh's *Zeta-function*, § 10.7; G. H. Hardy and J. E. Littlewood, *Math. Zeitschrift* 10 (1921), 283 – 317. For A. Selberg's improvement see *Skr. Norske Vid.*

Akad. Oslo (1942), No. 10; and the exposition in Titchmarsh, loc. cit., §§ 10.9 – 10.22.

For information regarding the Riemann hypothesis, see C. L. Siegel, loc. cit. in the notes on § 1, and Titchmarsh's *Zeta-function*, Chs. XIV, XV.

§ 5. Hamburger proved Theorem 9 in *Math. Zeitschrift*, 10 (1921), 240 – 254. Siegel's proof appeared in *Math. Annalen*, 86 (1922), 276 – 279; *Gesammelte Abhandlungen*, I, 154 – 156. It uses the absolute convergence of the series $\sum a_n n^{-s}$ only in the half-plane $\sigma > 2 - \varepsilon$, $\varepsilon > 0$. Another proof of Hamburger's theorem is given by E. Hecke, *Math. Zeitschrift*, 16 (1923), 301 – 7. For remarks on the connexion between the proofs of Siegel and of Hecke, see the paper by the author and Raghavan Narasimhan, *Annals of Math.* 74 (1961), 1 – 23, Lemmas 5 and 6. Hecke's general study of the correspondence between modular forms and Dirichlet series satisfying functional equations takes off from Hamburger's theorem. See his Lecture Notes on *Dirichlet series, modular functions, and quadratic forms*, Princeton University, 1938.

Siegel's *partial-fraction formula* (72) has led to further generalizations of Hamburger's theorem. See, for instance, S. Bochner and the author, *Annals of Math.* 63 (1956), 336 – 360; also the author and S. Mandelbrojt, *Annals of Math.* 66 (1957), 285 – 296. The author's proof of a generalization of Siegel's formula (72) using the Mellin transform of the beta-function, instead of the theta-relation, is reproduced by J. P. Kahane and S. Mandelbrojt, *Annales. Sci. l'École Norm. Sup.* 75 (1958), 65. See also Lemma 3 in the paper by the author and Raghavan Narasimhan, *Acta Arithmetica*, 6 (1961), 487 – 503.

Formula (62) is a consequence of the fact that e^{-x} and $\Gamma(s)$ are Mellin transforms. See Titchmarsh's *Fourier integrals*, §§ 1.5, 1.29, 2.1, 2.7, 3.17. It can be proved directly by contour integration. See, for instance, Landau's *Vorlesungen*, I, Satz 231. Formula (70) can be proved by the method of Fourier transforms. See, for instance, S. Bochner's book, (loc. cit. in Notes on § 1), 57. It can also be obtained directly by differentiation with respect to b. See Titchmarsh's *Theory of functions*, 2nd edition (1939), 62.

For the Phragmén-Lindelöf principle, see, for instance, J. E. Littlewood's *Theory of functions* (loc. cit. in Notes on § 3), 107.

Chapter III

Littlewood's theorem and Weyl's method

§ 1. Zero-free region of ζ. We have proved in Chapter II, Theorem 4, that all the non-real zeros of Riemann's zeta-function $\zeta(s)$ lie in the critical strip $0 \leqslant \sigma \leqslant 1$, and are symmetrically situated about the lines $\sigma = \frac{1}{2}$ and $t=0$, where $\sigma = \operatorname{Re} s$, $t = \operatorname{Im} s$. We shall now prove a theorem of J. E. Littlewood that there exists a positive constant A such that the region

$$\sigma > 1 - \frac{A \log\log(|t|+3)}{\log(|t|+3)} \tag{1}$$

is free from any zero of ζ.

Following Landau, we shall deduce Littlewood's theorem from the estimate

$$\zeta(s) = O(\log^{A_1} t), \tag{2}$$

where A_1 is a positive constant, in the region $\sigma \geqslant 1 - \dfrac{(\log\log t)^2}{\log t}$, $t > t_1$. This estimate is a direct consequence of a theorem of G.H. Hardy and J. E. Littlewood that, as $t \to \infty$,

$$\zeta(s) = O\left(t^{4(1-\sigma)/\log \frac{1}{1-\sigma}} \frac{\log t}{\log\log t}\right), \tag{3}$$

uniformly for $\frac{63}{64} \leqslant \sigma < 1$. By continuity, (3) implies Hermann Weyl's result that

$$\zeta(1+it) = O\left(\frac{\log t}{\log\log t}\right), \tag{4}$$

as $t \to \infty$.

The proof of (3), like that of (4), depends on estimates of sums of the form $\sum\limits_{n=N}^{N'} n^{-it}$, which in turn depend on estimates of sums of the form

$\sum_{n=N}^{N'} e^{iP(n)}$, where P is a polynomial with real coefficients. We shall obtain such estimates by a method originated by Hermann Weyl and developed by Hardy and Littlewood.

Littlewood's theorem leads to an estimate of the error term in the prime number theorem. If $\pi(x)$ denotes the number of primes not exceeding x, the prime number theorem is the assertion that

$$\pi(x) \sim \frac{x}{\log x}, \quad \text{as} \quad x \to \infty . \tag{5}$$

If we define, for $x > 1$, the *logarithmic integral* of x, namely

$$\operatorname{li} x = \lim_{\delta \to 0} \left(\int_0^{1-\delta} + \int_{1+\delta}^{x} \right) \frac{du}{\log u}, \tag{6}$$

then

$$\operatorname{li} x = \operatorname{li} 2 + \int_2^x \frac{du}{\log u}. \tag{7}$$

Since

$$\int_2^x \frac{du}{\log u} = \frac{x}{\log x} + o\left(\frac{x}{\log x}\right), \quad x \geqslant 4,$$

an equivalent form of (5) is

$$\pi(x) \sim \operatorname{li} x, \quad \text{as} \quad x \to \infty . \tag{8}$$

The *error term* in this asymptotic relation is $\pi(x) - \operatorname{li} x$. We shall prove that

$$\pi(x) - \operatorname{li} x = O\left(x e^{-a\sqrt{\log x \log\log x}}\right) \tag{9}$$

for a positive, absolute constant a. Since the prime number theorem is equivalent to the assertion that

$$\psi(x) \sim x, \quad \text{as} \quad x \to \infty, \tag{10}$$

where ψ is Chebyshev's function, we prove along with (9) the corresponding result on ψ, namely

$$\psi(x) - x = O\left(x e^{-a\sqrt{\log x \log\log x}}\right). \tag{11}$$

It can be shown, on the other hand, that *if* the Riemann hypothesis is true, then

$$\pi(x) - \operatorname{li} x = O(x^{\frac{1}{2}} \log x), \tag{12}$$

and

$$\psi(x) - x = O(x^{\frac{1}{2}} \log^2 x). \tag{13}$$

§ 2. Weyl's inequality. The estimation of a general sum of the form $\sum_{n=N}^{N'} n^{-it}$ is reduced to that of a *Weyl sum* $\sum_{n=N}^{N'} e^{iP(n)}$, where P is a polynomial with real coefficients, by means of the following

LEMMA 1. *Let k, μ be positive integers, t a real number, $t \geqslant 1$, a and b positive integers, $b-a \geqslant 1$, and $\frac{b-a}{a} \leqslant \frac{1}{2} t^{-1/(k+1)}$. Let M be such that*

$$\left| \sum_{n=1}^{\mu} \exp\left\{ -it\left(\frac{n}{a} - \frac{1}{2}\frac{n^2}{a^2} + \cdots + \frac{(-1)^{k-1} n^k}{k a^k} \right) \right\} \right| \leqslant M, \tag{14}$$

for $\mu \leqslant b-a$. Then

$$\left| \sum_{n=a+1}^{b} e^{-it\log n} \right| < 4M. \tag{15}$$

PROOF. We have

$$\left| \sum_{n=a+1}^{b} e^{-it\log n} \right| = \left| \sum_{m=1}^{b-a} e^{-it\log(a+m)} \right|$$

$$= \left| \sum_{m=1}^{b-a} \exp\left\{ -it\left(\frac{m}{a} - \cdots + \frac{(-1)^{k-1} m^k}{k a^k} \right) - it\left(\frac{(-1)^k m^{k+1}}{(k+1) a^{k+1}} + \cdots \right) \right\} \right|,$$

since $\frac{m}{a} \leqslant \frac{b-a}{a} \leqslant \frac{1}{2}$. For $0 \leqslant x < 1$, we have the expansion

$$\exp\left\{ -it \sum_{\nu=k+1}^{\infty} (-1)^{\nu-1} \frac{x^\nu}{\nu} \right\} = \sum_{l=0}^{\infty} e_l(t) x^l,$$

so that

$$\exp\left\{ -it\left(\frac{(-1)^k m^{k+1}}{(k+1) a^{k+1}} + \cdots \right) \right\} = \sum_{\nu=0}^{\infty} e_\nu(t) \left(\frac{m}{a} \right)^\nu.$$

Hence

$$\left| \sum_{n=a+1}^{b} e^{-it\log n} \right| = \left| \sum_{m=1}^{b-a} \exp\left\{ -it\left(\frac{m}{a} - \cdots + \frac{(-1)^{k-1} m^k}{k a^k} \right) \right\} \sum_{\nu=0}^{\infty} e_\nu(t) \left(\frac{m}{a} \right)^\nu \right|$$

$$= \left| \sum_{\nu=0}^{\infty} \frac{e_\nu(t)}{a^\nu} \sum_{m=1}^{b-a} m^\nu \exp\left\{ -it\left(\frac{m}{a} - \cdots + \frac{(-1)^{k-1} m^k}{k a^k} \right) \right\} \right|.$$

If we use partial summation in the second sum, as well as the estimate (14), we get

$$\left|\sum_{n=a+1}^{b} e^{-it\log n}\right| \leqslant 2M \sum_{\nu=0}^{\infty} |e_\nu(t)| \left(\frac{b-a}{a}\right)^{\nu}$$

$$\leqslant 2M \exp\left\{t\left(\frac{(b-a)^{k+1}}{(k+1)a^{k+1}} + \cdots\right)\right\}$$

$$\leqslant 2M \exp\left\{\frac{t\left(\frac{b-a}{a}\right)^{k+1}}{1-\frac{b-a}{a}}\right\}.$$

Now

$$\frac{t\left(\frac{b-a}{a}\right)^{k+1}}{1-\frac{b-a}{a}} \leqslant \frac{t\left(\frac{1}{2}t^{-\frac{1}{(k+1)}}\right)^{k+1}}{1-\frac{1}{2}t^{-1/(k+1)}} \leqslant \frac{2^{-(k+1)}}{\frac{1}{2}} = \frac{1}{2^k},$$

and $e^{1/(2^k)}<2$. Hence we obtain the lemma.

Now let

$$S = \sum_{m=1}^{\mu} e^{2\pi i P(m)}, \tag{16}$$

where μ is a positive integer and P a polynomial of degree k with real coefficients. Trivially we have

$$|S| \leqslant \mu .$$

But we can obtain a better inequality.

If we consider the case $P(m)=\alpha m$, α real, then obviously we have, for α non-integral,

$$|S| = \left|\frac{1-e^{2\pi i\alpha\mu}}{1-e^{2\pi i\alpha}}\right| \leqslant |\operatorname{cosec}\pi\alpha|,$$

so that

$$|S| \leqslant \min(\mu, |\operatorname{cosec}\pi\alpha|). \tag{17}$$

If we consider the case $P(m)=\alpha m^2+\beta m$, with α and β real, then

$$|S|^2 = \sum_{m=1}^{\mu} \sum_{m'=1}^{\mu} e^{2\pi i(\alpha m^2+\beta m-\alpha m'^2-\beta m')}.$$

On setting $m'=m-r$, this double sum takes the form

$$\sum_m \sum_r e^{2\pi i(2\alpha m r-\alpha r^2+\beta r)},$$

which, in absolute value, is

$$\leqslant \sum_{r=-\mu+1}^{\mu-1} \left| \sum_{m=\max(r+1,1)}^{\min(r+\mu,\mu)} e^{4\pi i \alpha m r} \right|,$$

the inner sum extending over at most μ consecutive integers. Hence, by (17), we have

$$\begin{aligned} |S|^2 &\leqslant \sum_{r=-\mu+1}^{\mu-1} \min(\mu, |\operatorname{cosec}(2\pi\alpha r)|) \\ &= \mu + 2\sum_{r=1}^{\mu-1} \min(\mu, |\operatorname{cosec}(2\pi\alpha r)|). \end{aligned} \tag{18}$$

Inequalities (17) and (18) give estimates of the sum in (16) when the degree of the polynomial P is 1 or 2. We shall now consider the case when the degree is arbitrary.

THEOREM 1 (WEYL'S INEQUALITY). *Let k be a positive integer, and*

$$P(m) = \alpha m^k + \alpha_{k-1} m^{k-1} + \cdots + \alpha_1 m,$$

where $\alpha, \alpha_1, \ldots, \alpha_{k-1}$ are real, and $\alpha \neq 0$. Let μ be a positive integer, and

$$S = \sum_{m=1}^{\mu} e^{2\pi i P(m)}.$$

Let $K = 2^{k-1}$, $k \geqslant 2$. Then we have

$$|S|^K \leqslant 2^{2K} \mu^{K-1} + 2^K \mu^{K-k} \sum_{r_1, \ldots, r_{k-1}} \min(\mu, |\operatorname{cosec}(\pi\alpha k! R)|), \tag{19}$$

where $R = r_1 \cdot r_2 \cdot \cdots \cdot r_{k-1}$, and each r_j varies from 1 to $\mu - 1$.

If $k = 1$, (19) holds with the sum replaced by the single term $\min(\mu, |\operatorname{cosec} \pi\alpha|)$.

PROOF. If $k = 1$, the conclusion results from (17). So we consider only the case $k \geqslant 2$.

We have

$$\begin{aligned} |S|^2 &= \sum_m \sum_{m'} e^{2\pi i \{P(m) - P(m')\}} \\ &= \sum_m \sum_{r_1} e^{2\pi i \{P(m) - P(m - r_1)\}}, \qquad m' = m - r_1 \\ &\leqslant \sum_{r_1 = -\mu+1}^{\mu-1} |S_1|, \end{aligned}$$

where

$$
\begin{aligned}
S_1 = S_1(r_1) &= \sum_{m=\max(r_1+1,1)}^{\min(r_1+\mu,\mu)} e^{2\pi i\{P(m)-P(m-r_1)\}} \\
&= \sum_{m=\max(r_1+1,1)}^{\min(r_1+\mu,\mu)} e^{2\pi i\{\alpha k r_1 m^{k-1}+\cdots\}}.
\end{aligned} \tag{20}
$$

By Hölder's inequality,

$$
\begin{aligned}
|S|^2 &\leqslant \left(\sum_{r_1=-\mu+1}^{\mu-1} 1\right)^{1-2/K} \left(\sum_{r_1=-\mu+1}^{\mu-1} |S_1|^{K/2}\right)^{2/K} \\
&\leqslant (2\mu)^{1-2/K}\left(\mu^{K/2} + \sum_{r_1=-\mu+1}^{\mu-1}{}' |S_1|^{K/2}\right)^{2/K},
\end{aligned}
$$

where the dash indicates the omission of the term corresponding to $r_1=0$. Hence

$$
|S|^K \leqslant (2\mu)^{\frac{K}{2}-1}\left(\mu^{K/2} + \sum_{r_1=-\mu+1}^{\mu-1}{}' |S_1|^{K/2}\right). \tag{21}
$$

Now if (19) is true with $k-1$ in place of k, we shall show that it is true also for k. Since (18) shows that it is true for $k=2$, it follows that it is true for all $k \geqslant 2$.

If (19) is true with $k-1$ in place of k, we apply it to estimate S_1, and obtain

$$
\begin{aligned}
|S_1|^{K/2} &\leqslant 2^K \mu^{\frac{K}{2}-1} \\
&\quad + 2^{K/2} \mu^{\frac{1}{2}K-k+1} \sum_{r_2,\ldots,r_{k-1}} \min\left(\mu, |\operatorname{cosec}(r_1 \pi \alpha k (r_2 \ldots r_{k-1})(k-1)!)|\right).
\end{aligned}
$$

Hence (21) gives

$$
|S|^K \leqslant 2^{\frac{K}{2}-1} \mu^{K-1} + 2^{\frac{3K}{2}} \mu^{K-1} + 2^K \mu^{K-k} \sum_{r_1,\ldots,r_{k-1}} \min(\mu, |\operatorname{cosec}(\pi \alpha k!\, R)|),
$$

which implies (19).

Weyl's inequality can be used to estimate sums of the type $\sum_{n=a+1}^{b} n^{-it}$, as shown by the following

THEOREM 2. *Let $t>3$; a and b integers, such that $a<b\leqslant 2a$; k a positive integer; $K=2^{k-1}$. Then we have*

$$
\left|\sum_{n=a+1}^{b} n^{-it}\right| < c_1\left(a^{1-\frac{1}{K}} t^{\frac{1}{(k+1)K}} + a t^{-\frac{1}{(k+1)K}} \log^{\frac{k-1}{K}} a\right) \log^{1/K} t, \tag{22}
$$

where c_1 is a positive, absolute constant.

PROOF. The result is easily established if $a \leqslant 4t^{1/(k+1)}$, for

$$\left|\sum_{n=a+1}^{b} n^{-it}\right| \leqslant \sum_{n=a+1}^{b} 1 = b-a \leqslant a \leqslant a^{1-\frac{1}{K}} t^{\frac{1}{(k+1)K}} 4^{1/K}$$

$$< 4a^{1-\frac{1}{K}} t^{\frac{1}{(k+1)K}} \log^{1/K} t < B, \tag{23}$$

where B denotes the right-hand side of the inequality (22).

We have therefore only to consider the case $a > 4t^{1/(k+1)}$, which implies that $a > 4$.

If we set

$$\mu = [\tfrac{1}{2} a t^{-\frac{1}{(k+1)}}], \qquad N = \left[\frac{b-a}{\mu}\right], \tag{24}$$

then obviously we have $\mu \geqslant 2$, and

$$\tfrac{1}{4} a t^{-1/(k+1)} < \mu \leqslant \tfrac{1}{2} a t^{-1/(k+1)}, \tag{25}$$

$$0 \leqslant N \leqslant \frac{b-a}{\mu} \leqslant \frac{a}{\mu} < 4t^{1/(k+1)}, \tag{26}$$

$$\mu N \leqslant b-a < \mu N + \mu. \tag{27}$$

If $N=0$ or 1, then the theorem is again easy to prove, for $b-a<2\mu$ by (27), so that we have

$$\left|\sum_{n=a+1}^{b} n^{-it}\right| \leqslant b-a < 2\mu \leqslant a t^{-\frac{1}{(k+1)}} < a t^{-\frac{1}{(k+1)K}} \log^{\frac{k-1}{K}} a \cdot \log^{\frac{1}{K}} t < B.$$

It remains for us therefore to prove the theorem in the case $a > 4t^{1/(k+1)}$, $N \geqslant 2$. We can now write

$$\sum = \sum_{n=a+1}^{b} n^{-it} = \sum_{a+1}^{a+\mu} + \sum_{n=a+\mu+1}^{a+2\mu} + \cdots + \sum_{n=a+\mu N+1}^{b}$$

$$= \sum\nolimits_1 + \sum\nolimits_2 + \cdots + \sum\nolimits_{N+1}, \tag{28}$$

say, by splitting up the range of summation.

If we set

$$v_m = a + m\mu, \qquad 0 \leqslant m \leqslant N+1, \tag{29}$$

then, for $1 \leqslant m \leqslant N+1$, we have, by (27), and because $N \geqslant 2$,

$$a < a+\mu = v_1 \leqslant v_m \leqslant v_{N+1} = a + (N+1)\mu \leqslant a+\mu+b-a$$

$$= b+\mu \leqslant 2a+\mu \tag{30}$$

$$\leqslant 2a + \frac{b-a}{2} \leqslant \frac{5a}{2} < 3a,$$

while

$$\frac{\mu}{v_m} < \frac{\mu}{a} \leqslant \frac{1}{2} t^{-\frac{1}{(k+1)}}, \qquad t > 3. \tag{31}$$

We can use Lemma 1 to estimate $\sum_m$, $m=1,2,\ldots,N+1$. If M denotes the maximum, for $1\leqslant\mu'\leqslant\mu$, of the modulus of the sum

$$S_m=\sum_{n=1}^{\mu'}\exp\left\{-it\left(\frac{n}{a+m\mu}-\frac{1}{2}\,\frac{n^2}{(a+m\mu)^2}+\cdots+\frac{(-1)^{k-1}n^k}{k(a+m\mu)^k}\right)\right\},$$

then, by Lemma 1, we have

$$|\textstyle\sum_{m+1}|<4M,\qquad m=0,1,2,\ldots,N. \tag{32}$$

But Theorem 1 gives us the estimate

$$|S_m|\leqslant 2^2\mu^{1-\frac{1}{K}}+2\mu^{1-\frac{k}{K}}\left\{\sum_{r_1,\ldots,r_{k-1}}\min\left(\mu,\left|\operatorname{cosec}\frac{t(k-1)!R}{2(a+m\mu)^k}\right|\right)\right\}^{\frac{1}{K}},$$

since $(A+B)^{\frac{1}{K}}\leqslant A^{\frac{1}{K}}+B^{\frac{1}{K}}$ for $A,B\geqslant 0$. Hence, by (28) and (32), we have

$$|\textstyle\sum|\leqslant 2^4(N+1)\mu^{1-\frac{1}{K}}$$
$$+2^3\mu^{1-\frac{k}{K}}\sum_{m=0}^{N}\left\{\sum_{r_1,\ldots,r_{k-1}}\min\left(\mu,\left|\operatorname{cosec}\frac{t(k-1)!R}{2(a+m\mu)^k}\right|\right)\right\}^{\frac{1}{K}}.$$

By Hölder's inequality, this gives

$$|\textstyle\sum|\leqslant 2^4(N+1)\mu^{1-\frac{1}{K}}$$
$$+2^3\mu^{1-\frac{k}{K}}(N+1)^{1-\frac{1}{K}}\left\{\sum_{m=0}^{N}\sum_{r_1,\ldots,r_{k-1}}\min\left(\mu,\left|\operatorname{cosec}\frac{t(k-1)!R}{2(a+m\mu)^k}\right|\right)\right\}^{\frac{1}{K}}, \tag{33}$$

where $R=r_1\cdot r_2\ldots r_{k-1}$. To estimate the multiple sum in the last term of (33), we set

$$\theta_m=\theta_m(t,k,\mu,R)=\frac{t(k-1)!R}{2(a+m\mu)^k}=\frac{t(k-1)!R}{2v_m^k},\quad\text{(cf. (29))} \tag{34}$$

and seek first to estimate the sum

$$\sum_{m=0}^{N}\min(\mu,|\operatorname{cosec}\theta_m|).$$

We note that, for fixed R, θ_m decreases as m increases, and

$$\begin{aligned}\theta_m-\theta_{m+1} &= \frac{tk!R}{2k}\left(\frac{1}{v_m^k}-\frac{1}{v_{m+1}^k}\right) > \frac{tk!R}{2k}\cdot\frac{k(v_{m+1}-v_m)}{v_{m+1}^{k+1}}\\ &> \frac{tk!R}{2k}\,\frac{k\mu}{(3a)^{k+1}} \quad \text{(by (30))}\\ &> \frac{tk!R}{2}\cdot\frac{at^{-\frac{1}{(k+1)}}}{4(3a)^{k+1}} \quad \text{(by (25))}\\ &= \frac{k!\,R\,t^{\frac{k}{(k+1)}}a^{-k}}{2^3\,3^{k+1}} = \Delta, \end{aligned} \tag{35}$$

say, where Δ is independent of m. Further

$$0<\theta_1 = \frac{tk!R}{2k(a+\mu)^k} < \frac{tk!R}{a^k}. \tag{36}$$

Let us consider all the values of m, say $m_1 \leqslant m \leqslant m_{l+1}$, where $m_{l+1} = m_1+l$, l being a non-negative integer, for which θ_m lies in an interval of the form $[g\pi, (g+\frac{1}{2})\pi]$, where g is an integer. Since, for fixed R, θ_m decreases as m increases, it follows that if such an interval contains θ_{m_0} and $\theta_{m'}$, where $m'>m_0$, then it contains all θ_m such that $m_0 \leqslant m \leqslant m'$. If $\theta_m = g\pi+\alpha$, $0\leqslant\alpha\leqslant\frac{1}{2}\pi$, then $|\sin\theta_m| = \sin\alpha \geqslant \frac{2\alpha}{\pi}$, so that by (35),

$$|\sin\theta_m| \geqslant (m_{l+1}-m)\cdot\frac{2}{\pi}\cdot\Delta, \quad \text{for} \quad m_1\leqslant m\leqslant m_{l+1}. \tag{37}$$

For $m=m_{l+1}$, it is possible that $|\sin\theta_m| = |\sin g\pi| = 0$. Hence we have

$$\begin{aligned}\sum_{\theta_m\in[g\pi,(g+\frac{1}{2})\pi]} \min(\mu, |\operatorname{cosec}\theta_m|) &\leqslant \mu + \frac{1}{2}\pi\sum_{m=1}^{N}\frac{1}{m\Delta}\\ &\leqslant \mu + \frac{1}{2}\pi\left(\frac{1+\log N}{\Delta}\right) \leqslant \mu + 2\left(\frac{1+\log N}{\Delta}\right)\\ &< \mu + \frac{2}{\Delta}(1+\log(4t^{\frac{1}{2}})), \quad \text{by (26)}\\ &< \mu + \frac{2}{\Delta}(3+\tfrac{1}{2}\log t)\\ &< \mu + \frac{7}{\Delta}\log t, \quad \text{since} \quad t>3.\end{aligned}$$

The same reasoning applies to the interval $[(g-\frac{1}{2})\pi, g\pi]$, so that

$$\sum_{\theta_m \in [(g-\frac{1}{2})\pi, g\pi]} \min(\mu, |\operatorname{cosec}\theta_m|) < \mu + \frac{7}{\Delta}\log t. \tag{38}$$

Now, by (35) and (25), we have

$$\mu\Delta = \frac{\mu k!\, R\, t^{k/(k+1)} a^{-k}}{2^3\, 3^{k+1}} < \frac{\mu^k k!\, t^{k/(k+1)} a^{-k}}{2^{k+4}}$$
$$< k!(\mu t^{1/(k+1)} a^{-1})^k < k!\,.$$

Hence

$$\sum_{\theta_m \in I} \min(\mu, |\operatorname{cosec}\theta_m|) < \left(\frac{k!}{\Delta} + \frac{7}{\Delta}\right)\log t \leqslant \frac{8k!}{\Delta}\log t$$
$$\leqslant 2^{10k} t^{-k/(k+1)} \log t \cdot a^k \cdot \frac{1}{R}, \tag{39}$$

where I stands for either of the intervals $[(g-\frac{1}{2})\pi, g\pi]$, $[g\pi, (g+\frac{1}{2})\pi]$. By (36) the number of such intervals I to which some θ_m, $0 \leqslant m \leqslant N$, belongs is *at most*

$$1 + \frac{2}{\pi} t k!\, R a^{-k} < 1 + t k!\, R a^{-k}. \tag{40}$$

From (39) and (40) we obtain the estimate

$$\sum_{m=0}^{N} \min(\mu, |\operatorname{cosec}\theta_m|) < (1 + t k!\, R a^{-k})\, 2^{10k} t^{-k/(k+1)} a^k \cdot \frac{1}{R} \cdot \log t$$
$$= 2^{10k} \cdot t^{-k/(k+1)} \cdot a^k \cdot \frac{1}{R} \log t + 2^{10k} k!\, t^{1/(k+1)} \log t. \tag{41}$$

Now

$$\sum_{r_1=1}^{\mu-1} \cdots \sum_{r_{k-1}=1}^{\mu-1} \frac{1}{r_1 \cdot r_2 \ldots r_{k-1}} = \left(\sum_{n=1}^{\mu-1} \frac{1}{n}\right)^{k-1} \leqslant (1 + \log(\mu-1))^{k-1}$$
$$\leqslant (1 + \log a)^{k-1} < (2 \log a)^{k-1}, \tag{42}$$

since $a \geqslant 2\mu$, $\mu \geqslant 2$ and $a > 4$, while

$$\sum_{r_1=1}^{\mu-1} \cdots \sum_{r_{k-1}=1}^{\mu-1} 1 = (\mu-1)^{k-1} < \mu^{k-1} \leqslant a^{k-1} t^{-\frac{k-1}{k+1}}, \quad \text{by (25)}. \tag{43}$$

Now

$$\sum_{m=0}^{N} \sum_{r_1, \ldots, r_{k-1}} \min(\mu, |\operatorname{cosec}\theta_m|) = \sum_{r_1, \ldots, r_{k-1}} \sum_{m=0}^{N} \min(\mu, |\operatorname{cosec}\theta_m|),$$

and if we use (41), (42) and (43) to estimate the last sum, we get

$$\sum_{m=0}^{N} \sum_{r_1,\ldots,r_{k-1}} \min(\mu, |\operatorname{cosec}\theta_m|)$$

$$< (2\log a)^{k-1} \cdot 2^{10k} t^{-\frac{k}{k+1}} a^k \log t + 2^{10k} k!\, t^{\frac{2-k}{k+1}} \cdot \log t \cdot a^{k-1}$$

$$< 2^{11K} \log^{k-1} a \cdot t^{-\frac{k}{k+1}} a^k \log t + 2^{11K} t^{\frac{2-k}{k+1}} \log t \cdot a^{k-1},$$

since $k \leqslant K$, and $k! \leqslant \prod_{n=2}^{k} 2^{2^{n-2}} < 2^{2^{k-1}} = 2^K$ for $k \geqslant 2$. Hence

$$\left(\sum_{m=0}^{N} \sum_{r_1,\ldots,r_{k-1}} \min(\mu, |\operatorname{cosec}\theta_m|)\right)^{\frac{1}{K}}$$

$$< 2^{11}\left(t^{-\frac{k}{(k+1)K}} a^{\frac{k}{K}} \log^{\frac{k-1}{K}} a + t^{\frac{2-k}{(k+1)K}} a^{\frac{k-1}{K}}\right) \log^{1/K} t = V, \text{ say}. \tag{44}$$

If we use this estimate in (33), we get

$$\left|\sum_{n=a+1}^{b} n^{-it}\right| < 2^4 (N+1) \mu^{1-\frac{1}{K}} + 2^3 (N+1)^{1-\frac{1}{K}} \mu^{1-\frac{k}{K}} V. \tag{45}$$

But by (25) and (26) we have

$$2^4 (N+1) \mu^{1-\frac{1}{K}} < 2^4 (N+1) a^{1-\frac{1}{K}} t^{-\frac{1}{k+1}\left(1-\frac{1}{K}\right)}$$

$$< 2^4 \left(4 t^{\frac{1}{k+1}} + 1\right) a^{1-\frac{1}{K}} t^{-\frac{1}{k+1} + \frac{1}{K(k+1)}} \tag{46}$$

$$< 2^7 a^{1-\frac{1}{K}} t^{\frac{1}{(k+1)K}} \log^{1/K} t,$$

and similarly

$$2^3 (N+1)^{1-\frac{1}{K}} \mu^{1-\frac{k}{K}} < 2^6 \left(t^{\frac{1}{k+1}}\right)^{1-\frac{1}{K}} \left(a t^{-\frac{1}{k+1}}\right)^{1-\frac{k}{K}} = 2^6 a^{1-\frac{k}{K}} t^{\frac{k}{(k+1)K} - \frac{1}{K(k+1)}}. \tag{47}$$

If we use (47), (46), and (44) in (45), we get the estimate

$$\left|\sum_{n=a+1}^{b} n^{-it}\right| < 2^7 a^{1-\frac{1}{K}} t^{\frac{1}{(k+1)K}} \log^{1/K} t$$

$$+ 2^{17}\left(a \log^{\frac{k-1}{K}} a \cdot t^{-\frac{1}{(k+1)K}} + a^{1-\frac{1}{K}} t^{\frac{1}{(k+1)K}}\right) \log^{1/K} t$$

$$< 2^{18}\left(a \log^{\frac{k-1}{K}} a \cdot t^{-\frac{1}{(k+1)K}} + a^{1-\frac{1}{K}} t^{\frac{1}{(k+1)K}}\right) \log^{1/K} t,$$

which proves (22), and hence the theorem.

§ 3. Some results of Hardy and Littlewood and of Weyl. We shall use Theorem 2 to prove the following

THEOREM 3 (HARDY-LITTLEWOOD, WEYL). *For* $t \geqslant 3$, *we have*

$$\zeta(s) = O\left(t^{4(1-\sigma)/\log\frac{1}{1-\sigma}} \frac{\log t}{\log\log t}\right), \tag{48}$$

uniformly for $\frac{63}{64} \leqslant \sigma < 1$. *By continuity,* (48) *implies that*

$$\zeta(1+it) = O\left(\frac{\log t}{\log\log t}\right). \tag{49}$$

Further

$$\zeta(s) = O(\log^A t), \qquad A > 1, \tag{50}$$

for $\sigma \geqslant 1 - \dfrac{(\log\log t)^2}{\log t}$, $t \geqslant 3$.

PROOF. As a first step, we prove that

$$\zeta(s) = \sum_{n=1}^{[t^2]} n^{-s} + O(1), \tag{51}$$

uniformly for $\frac{31}{32} \leqslant \sigma < 1$, $t > 3$. This follows from the relation

$$\zeta(s) - \sum_{n=1}^{N} n^{-s} = -s \int_N^\infty \frac{u - [u]}{u^{s+1}}\, du + \frac{1}{(s-1)N^{s-1}}, \tag{52}$$

which is valid for $\sigma > 0$, and any integer $N \geqslant 1$. (Ch. II, (16), (17)). If we choose $N = [t^2]$, $t > 3$, and $\frac{31}{32} \leqslant \sigma < 1$, then we have

$$\begin{aligned}\zeta(s) &= \sum_{n=1}^{[t^2]} n^{-s} + O\left(t \int_N^\infty \frac{du}{u^{\frac{31}{32}+1}}\right) + O(N^{1-\sigma} t^{-1}) \\ &= \sum_{n=1}^{[t^2]} n^{-s} + O(tN^{-31/32}) + O(tN^{-31/32}), \\ &= \sum_{n=1}^{[t^2]} n^{-s} + O(1),\end{aligned}$$

the constant involved in the O's being absolute. This proves (51).

As a second step, we prove that

$$\zeta(s) = \sum_{n=1}^{[t^{1/2}]} n^{-s} + O(1), \tag{53}$$

uniformly for $\frac{31}{32}\leqslant\sigma<1$, $t>3$. This will follow from (51) if we show that

$$\Bigl|\sum_{t^{1/2}<n\leqslant t^2} n^{-s}\Bigr|=O(1), \quad \text{for} \quad \tfrac{31}{32}\leqslant\sigma<1, \quad t>3. \tag{54}$$

We split this sum as

$$\sum_{t^{1/2}<n\leqslant t^2} n^{-s}=\sum_{n=[t^{1/2}]+1}^{2[t^{1/2}]}+\sum_{n=2[t^{1/2}]+1}^{4[t^{1/2}]}+\cdots+\sum_{\frac{1}{2}t^2<2^r[t^{1/2}]<n\leqslant t^2},$$

the number of such sums being $O(\log t)$; and if h is a non-negative integer, such that $2^{h+1}[t^{\frac{1}{2}}]\leqslant t^2$, then

$$\sum_{n=2^h[t^{1/2}]+1}^{2^{h+1}[t^{1/2}]} n^{-s}=O(2^{-\sigma h}t^{-\sigma/2}M),$$

where

$$M=\sup_{2^h[t^{1/2}]+1\leqslant\mu\leqslant 2^{h+1}[t^{1/2}]}\Bigl|\sum_{2^h[t^{1/2}]+1}^{\mu} n^{-it}\Bigr|.$$

If we now appeal to Theorem 2 with $k=2=K$, we get

$$\begin{aligned}\Bigl|\sum_{2^h[t^{1/2}]+1}^{2^{h+1}[t^{1/2}]} n^{-s}\Bigr| &= O(2^{-\sigma h}t^{-\sigma/2})\{O(2^{h/2}\cdot t^{\frac{1}{4}}\cdot t^{\frac{1}{6}}\cdot\log^{\frac{1}{2}}t)+O(2^h\cdot t^{\frac{1}{2}}\cdot t^{-\frac{1}{6}}\log t)\}\\ &= O(t^{-\frac{15}{64}}\cdot t^{\frac{1}{6}}\cdot\log^{\frac{1}{2}}t)+O((2^h t^{\frac{1}{2}})^{1-\sigma}\cdot t^{-\frac{1}{6}}\cdot\log t)\\ &= O\left(\frac{1}{\log t}\right)+O(t^{2(1-\sigma)}\cdot t^{-\frac{1}{6}}\log t)\\ &= O\left(\frac{1}{\log t}\right).\end{aligned}$$

Since the number of such sums is $O(\log t)$, we have proved (54).

As a third step, we prove that

$$\zeta(s)=\sum_{1\leqslant n\leqslant t^{2/r}} n^{-s}+O(1), \tag{55}$$

uniformly for $1-\dfrac{1}{R}\leqslant\sigma<1$, where $R=2^{r-1}$, and r is an integer such that $6\leqslant r\leqslant\log\log t$. This will follow from (54) if we prove that

$$\sum_{t^{2/r}<n\leqslant t^{1/2}} n^{-s}=O(1), \tag{56}$$

in the given range of σ and of t. We shall prove (56) by proving that

$$\sum_{t^{\frac{2}{(k+2)}}<n\leqslant t^{\frac{2}{(k+1)}}} n^{-s}=O\left(t^{-\frac{1}{8kK}}\log^2 t\right), \tag{57}$$

for $k \geqslant 3$, $1-\frac{1}{4K} \leqslant \sigma < 1$, $t > 3$, k integral, $K = 2^{k-1}$. The O is uniform in k. This sum can be split up into

$$\sum_{n=\left[t^{\frac{2}{k+2}}\right]+1}^{2\left[t^{\frac{2}{k+2}}\right]} + \sum_{2\left[t^{\frac{2}{k+2}}\right]+1}^{4\left[t^{\frac{2}{k+2}}\right]} + \cdots + \sum_{2^h\left[t^{\frac{2}{k+2}}\right]+1}^{2^{h+1}\left[t^{\frac{2}{k+2}}\right]} + \cdots,$$

the number of such sums being $O(\log t)$, since

$$t^{\frac{2}{k+2}} \cdot 2^{\log t} > 2^{\log t} = t^{\log 2} > t^{2/(k+1)},$$

for $k \geqslant 3$, $t > 3$. We estimate these sums again with the help of Theorem 2, and obtain

$$\sum_{n=2^h[t^{2/(k+2)}]+1}^{2^{h+1}[t^{2/(k+2)}]} n^{-s} = O\left(\left(2^h t^{2/(k+2)}\right)^{-\sigma+1-\frac{1}{K}} t^{\frac{1}{(k+1)K}} \log t\right)$$

$$+ O\left(\left(2^h t^{2/(k+2)}\right)^{-\sigma+1} t^{-\frac{1}{(k+1)K}} \log t\right)$$

$$= O\left(t^{-\frac{3}{2(k+2)K}+\frac{1}{(k+1)K}} \log t\right) + O\left(t^{\frac{1}{2(k+1)K}-\frac{1}{(k+1)K}} \log t\right)$$

$$= O\left(t^{-\frac{1}{8kK}} \log t\right), \tag{58}$$

since

$$\frac{-3(k+1)+2(k+2)}{2(k+2)(k+1)K} = \frac{-k+1}{2(k+2)(k+1)K} = -\frac{1}{2(k+2)K} \cdot \frac{k-1}{k+1} < -\frac{1}{4kK} \cdot \frac{1}{2},$$

and

$$-\frac{1}{2(k+1)K} \leqslant -\frac{1}{8kK}.$$

Since there are $O(\log t)$ terms, each of which is of the order given by (58), we have proved (57).

To deduce (56) from (57), we note that if r is an integer such that $6 \leqslant r \leqslant \log\log t$, $R = 2^{r-1}$, and $1-\frac{1}{R} \leqslant \sigma < 1$, and $3 \leqslant k \leqslant r-2$, then $2^k \leqslant 2^{r-2}$, which implies that $4K = 2^2 \cdot 2^{k-1} = 2^{k+1} \leqslant 2^{r-1} = R$. Hence

$$\sigma \geqslant 1-\frac{1}{R} \geqslant 1-\frac{1}{4K},$$

and

$$Rr \geqslant 4Kr \geqslant 4K(k+2) \geqslant Kk,$$

and (57) gives

$$\sum_{t^{2/(k+2)}<n\leqslant t^{2/(k+1)}} n^{-s}=O\left(t^{-\frac{1}{8kK}}\log^2 t\right)=O\left(t^{-\frac{1}{8Rr}}\log^2 t\right)$$
$$=O(t^{-\frac{1}{8}(\log\log t)^{-1}2^{1-\log\log t}}\log^2 t),$$

since $Rr=2^{r-1}r\leqslant 2^{\log\log t-1}\log\log t$. But $2^{\log t}=t^{\log 2}$, so that $2^{\log\log t}=(\log t)^{\log 2}$, and $1>\log 2>\frac{1}{2}$. Hence

$$\sum_{t^{2/(k+2)}<n\leqslant t^{2/(k+1)}} n^{-s}=O\,(e^{-(\log t)^{1-\log 2}/(4\log\log t)}\log^2 t)=O\left(\frac{1}{\log t}\right),$$

which implies that

$$\sum_{t^{2/r}<n\leqslant t^{1/2}} n^{-s}=\sum_{k=3}^{r-2}\;\sum_{t^{2/(k+2)}<n\leqslant t^{2/(k+1)}} n^{-s}=O\left(\frac{r}{\log t}\right)=O\left(\frac{\log\log t}{\log t}\right)=O(1),$$

for $6\leqslant r\leqslant\log\log t$, so that (56) is proved, and hence the third step, namely (55), is completed.

To obtain the theorem from (55), we set

$$r=\text{integral part of }\min\left(\frac{\log\dfrac{1}{1-\sigma}}{\log 2},\ \log\log t\right),$$

where $t\geqslant e^{e^6}$. Since, by hypothesis, we have $\frac{63}{64}\leqslant\sigma<1$, it follows that $\dfrac{1}{1-\sigma}\geqslant 64=2^6$, so that $6\leqslant r\leqslant\log\log t$. Further, by the definition of r, we have $\sigma\geqslant 1-\dfrac{1}{2^r}>1-\dfrac{1}{R}$, where $R=2^{r-1}$. Hence we can use (55), and obtain

$$\zeta(s)=O\Big(\sum_{1\leqslant n\leqslant t^{2/r}} n^{-\sigma}\Big)+O(1). \tag{59}$$

We now consider two cases, according as $1-\sigma\leqslant\dfrac{\log\log t}{\log t}$, or $1-\sigma>\dfrac{\log\log t}{\log t}$.

Case (i). If $1-\sigma\leqslant\dfrac{\log\log t}{\log t}$, then for a sufficiently large t, we have

$$r\geqslant\left[\min\left(\frac{\log\log t-\log\log\log t}{\log 2},\ \log\log t\right)\right]>\frac{\log\log t}{2}.$$

Hence

$$\sum_{n=1}^{[t^{2/r}]} n^{-\sigma} = \sum_{n=1}^{[t^{2/r}]} \frac{n^{1-\sigma}}{n} \leqslant t^{\frac{2}{r}(1-\sigma)} \sum_{n=1}^{[t^{2/r}]} \frac{1}{n} < e^{\log t \cdot \frac{4}{\log\log t} \cdot \frac{\log\log t}{\log t}} \cdot 2\log(t^{2/r})$$

$$= O\left(\frac{\log t}{r}\right) = O\left(\frac{\log t}{\log\log t}\right) = O\left(t^{\frac{4(1-\sigma)}{\log\frac{1}{1-\sigma}}} \frac{\log t}{\log\log t}\right). \tag{60}$$

CASE (ii). If $1-\sigma > \dfrac{\log\log t}{\log t}$, then $\log\dfrac{1}{1-\sigma} < \log\log t$, and

$$r \geqslant \left[\log\frac{1}{1-\sigma}\right] > \frac{1}{2}\log\frac{1}{1-\sigma},$$

and

$$\sum_{n=1}^{[t^{2/r}]} n^{-\sigma} < \int_0^{t^{2/r}} \frac{du}{u^\sigma} = \frac{t^{(2/r)(1-\sigma)}}{1-\sigma} = O\left(t^{\frac{4(1-\sigma)}{\log\frac{1}{1-\sigma}}} \frac{\log t}{\log\log t}\right). \tag{61}$$

From (60) and (61) we obtain the conclusion (48) of the theorem, and (49) follows from it by continuity.

To prove (50), we have only to observe that if $1>\sigma\geqslant 1-\dfrac{(\log\log t)^2}{\log t}$, and t is sufficiently large, then by (48), we have uniformly

$$\zeta(s) = O\left(\exp\left\{4\log t\cdot\frac{(\log\log t)^2}{\log t}\Big/\log\frac{\log t}{(\log\log t)^2}\right\}\frac{\log t}{\log\log t}\right)$$

$$= O\left(\exp(A_1\log\log t)\cdot\frac{\log t}{\log\log t}\right) = O(\log^A t), \qquad A>1, \qquad t>t_1.$$

Because $\zeta(s)=O(\log t)$ for $\sigma\geqslant 1$, $t\geqslant 2$ by (14) of Chapter II, (50) follows.

§ 4. Littlewood's theorem.

§ 4. Littlewood's theorem. Our purpose is to deduce from Theorem 3 the existence of a region of the form $\sigma > 1 - \dfrac{A\log\log t}{\log t}$, $t>t_3$, which is free from zeros of ζ. We need the following function-theoretical

LEMMA 2 (LANDAU). *Let $r>0$, and f a function regular for $|s-s_0|\leqslant r$, $f(s_0)\neq 0$, M real,*

$$\left|\frac{f(s)}{f(s_0)}\right| < e^M, \quad \textit{for} \quad |s-s_0|\leqslant r,$$

so that $M>0$. Let $f(s)\neq 0$ in the semi-circle $|s-s_0|\leqslant r$, $\operatorname{Re}s>\operatorname{Re}s_0$.

Then we have

$$-\mathrm{Re}\left(\frac{f'(s_0)}{f(s_0)}\right) < \frac{4M}{r}. \tag{62}$$

If we further suppose that there exists a zero ρ_0 of f on the segment between $s_0-\frac{1}{2}r$ and s_0 (exclusive of the end-points), then

$$-\mathrm{Re}\left(\frac{f'(s_0)}{f(s_0)}\right) < \frac{4M}{r} - \frac{1}{s_0-\rho_0}. \tag{63}$$

PROOF (LANDAU). We may suppose, without loss of generality, that $f(s_0)=1$, for otherwise we can start with the function $\dfrac{f(s)}{f(s_0)} = f_1(s)$. We may also suppose that $s_0=0$, for otherwise we have only to consider $f(s-s_0)=f_2(s)$.

Thus, by hypothesis, $f(s)$ is regular for $|s|\leqslant r$, and $|f(s)|<e^M$, $M>0$, while $f(0)=1$, and $f(s)\neq 0$ for $|s|\leqslant r$, $\sigma>0$. We have to show that $-\mathrm{Re}\, f'(0) < \dfrac{4M}{r}$ to prove (62), and that $-\mathrm{Re}\, f'(0) < \dfrac{4M}{r} + \dfrac{1}{\rho_0}$ to prove (63).

Let ρ run through all the zeros of f in $|s|\leqslant\frac{1}{2}r$. Then the function

$$g(s) = \frac{f(s)}{\prod_\rho \left(1-\dfrac{s}{\rho}\right)}$$

is regular for $|s|\leqslant r$. Further we have, for $|s|=r$,

$$\left|1-\frac{s}{\rho}\right| \geqslant \left|\frac{s}{\rho}\right| - 1 = \frac{r}{|\rho|} - 1 \geqslant 1,$$

so that

$$|g(s)| \leqslant |f(s)| < e^M.$$

Since $g(0)=1$, and $g(s)$ has no zeros for $|s|\leqslant\frac{1}{2}r$, we can write $g(s)=e^{G(s)}$, where $G(s)$ is regular for $|s|\leqslant\frac{1}{2}r$, $\mathrm{Re}\, G(s)<M$, and $G(0)=0$. Hence by the Borel-Carathéodory inequality (Ch. II, Lemma B, (36)), we have $|G'(0)| < \dfrac{4M}{r}$. Therefore

$$\left|f'(0) + \sum_\rho \frac{1}{\rho}\right| = \left|\frac{g'(0)}{g(0)}\right| = |G'(0)| < \frac{4M}{r},$$

which implies that

$$-\operatorname{Re} f'(0) - \sum_\rho \operatorname{Re}\frac{1}{\rho} < \frac{4M}{r},$$

or

$$-\operatorname{Re} f'(0) < \frac{4M}{r} + \sum_\rho \operatorname{Re}\frac{1}{\rho}. \tag{64}$$

By hypothesis, we have $\rho \neq 0$, and $\operatorname{Re}\rho \leqslant 0$, for every ρ. Therefore $\operatorname{Re}\frac{1}{\rho} \leqslant 0$. If we use this fact in (64), we obtain $-\operatorname{Re} f'(0) < \frac{4M}{r}$, hence (62). If we further suppose that there exists a $\rho = \rho_0$ such that $-\frac{1}{2}r < \rho_0 < 0$, and retain only the corresponding term in (64), we get $-\operatorname{Re} f'(0) < \frac{4M}{r} + \frac{1}{\rho_0}$, hence (63).

Lemma 3 (Landau). *Let $\varphi(t)$ and $\frac{1}{\theta(t)}$ be positive, non-decreasing functions of t for $t \geqslant t_0$. Let $\theta(t) \leqslant 1$, $\varphi(t) \to \infty$ as $t \to \infty$, and*

$$\frac{\varphi(t)}{\theta(t)} = o(e^{\varphi(t)}). \tag{65}$$

Let

$$\zeta(s) = O(e^{\varphi(t)}), \tag{66}$$

in the region

$$1 - \theta(t) \leqslant \sigma \leqslant 2, \qquad t \geqslant t_0. \tag{67}$$

Then there exists a positive constant A_1, such that ζ has no zeros in the region $t \geqslant t_1$,

$$\sigma \geqslant 1 - A_1 \frac{\theta(2t+1)}{\varphi(2t+1)}. \tag{68}$$

Proof (Titchmarsh). Let $\beta + i\gamma$ be a zero of ζ in the upper half-plane. Let σ_0 be such that

$$1 + e^{-\varphi(2\gamma+1)} \leqslant \sigma_0 \leqslant 2.$$

Let $s_0 = \sigma_0 + i\gamma$, $s_0' = \sigma_0 + 2i\gamma$, $r = \theta(2\gamma+1)$, $\gamma > t_0 + 1$. Then the circles $|s - s_0| \leqslant r$ and $|s - s_0'| \leqslant r$ both lie in the region $\sigma \geqslant 1 - \theta(t)$, $t \geqslant t_0$, for

$$\begin{aligned}\sigma_0 - r = \sigma_0 - \theta(2\gamma+1) \geqslant 1 + e^{-\varphi(2\gamma+1)} - \theta(2\gamma+1) \geqslant 1 - \theta(2\gamma+1)\\ \geqslant 1 - \theta(\gamma+1).\end{aligned}$$

Since $\dfrac{1}{\zeta(s)}=\sum_{n=1}^{\infty}\mu(n)n^{-s}$, and $|\mu(n)|\leqslant 1$, we have, for $1<\sigma\leqslant 2$,

$$\left|\frac{1}{\zeta(s)}\right|\leqslant\zeta(\sigma)<\frac{A}{\sigma-1},\qquad A>0.$$

Hence

$$\left|\frac{1}{\zeta(s_0)}\right|<\frac{A}{\sigma_0-1}\leqslant A\,e^{\varphi(2\gamma+1)}.$$

Similarly

$$\left|\frac{1}{\zeta(s_0')}\right|<A\cdot e^{\varphi(2\gamma+1)}.$$

Since $\zeta(s)=O(e^{\varphi(t)})$ for $1-\theta(t)\leqslant\sigma\leqslant 2$, there exists a constant $A_2>0$, such that

$$\left|\frac{\zeta(s)}{\zeta(s_0)}\right|<e^{A_2\varphi(2\gamma+1)},\qquad |s-s_0|\leqslant r,$$

and

$$\left|\frac{\zeta(s)}{\zeta(s_0')}\right|<e^{A_2\varphi(2\gamma+1)},\qquad |s-s_0'|\leqslant r.$$

By Lemma 2, with $M=A_2\varphi(2\gamma+1)$, we get

$$-\operatorname{Re}\frac{\zeta'(\sigma_0+2i\gamma)}{\zeta(\sigma_0+2i\gamma)}<A_3\frac{\varphi(2\gamma+1)}{\theta(2\gamma+1)},\tag{69}$$

and if $\beta>\sigma_0-\frac{1}{2}r$, then

$$-\operatorname{Re}\frac{\zeta'(\sigma_0+i\gamma)}{\zeta(\sigma_0+i\gamma)}<A_3\frac{\varphi(2\gamma+1)}{\theta(2\gamma+1)}-\frac{1}{\sigma_0-\beta}.\tag{70}$$

Further, as $\sigma_0\to 1+0$, we have

$$-\frac{\zeta'(\sigma_0)}{\zeta(\sigma_0)}\sim\frac{1}{\sigma_0-1},$$

so that

$$-\frac{\zeta'(\sigma_0)}{\zeta(\sigma_0)}<\frac{a}{\sigma_0-1},\tag{71}$$

where $a\to 1+0$ as $\sigma_0\to 1+0$. Since $3+4\cos\theta+\cos 2\theta=2(1+\cos\theta)^2\geqslant 0$, for real θ, and

$$\frac{\zeta'(s)}{\zeta(s)}=-\sum_{n=1}^{\infty}\frac{\Lambda(n)}{n^s},\qquad\sigma>1,$$

where $\Lambda(n)$ is von Mangoldt's function (Ch. I, (9)), we have

$$-\operatorname{Re}\left(3\frac{\zeta'}{\zeta}(\sigma)+4\frac{\zeta'}{\zeta}(\sigma+it)+\frac{\zeta'}{\zeta}(\sigma+2it)\right)\geqslant 0, \qquad \sigma>1.$$

Hence, by (71), (70), and (69), we have

$$\frac{3a}{\sigma_0-1}+5A_3\frac{\varphi(2\gamma+1)}{\theta(2\gamma+1)}-\frac{4}{\sigma_0-\beta}\geqslant 0,$$

or

$$\sigma_0-\beta\geqslant\left(\frac{3a}{4(\sigma_0-1)}+\frac{5A_3}{4}\frac{\varphi(2\gamma+1)}{\theta(2\gamma+1)}\right)^{-1},$$

or

$$1-\beta\geqslant\left(\frac{3a}{4(\sigma_0-1)}+\frac{5A_3}{4}\frac{\varphi(2\gamma+1)}{\theta(2\gamma+1)}\right)^{-1}-(\sigma_0-1)$$

$$=\frac{1-\dfrac{3a}{4}-\dfrac{5A_3}{4}\dfrac{\varphi(2\gamma+1)}{\theta(2\gamma+1)}(\sigma_0-1)}{\dfrac{3a}{4(\sigma_0-1)}+\dfrac{5A_3}{4}\dfrac{\varphi(2\gamma+1)}{\theta(2\gamma+1)}}. \tag{72}$$

If we take $a=\frac{5}{4}$, and $\sigma_0-1=\dfrac{1}{40A_3}\cdot\dfrac{\theta(2\gamma+1)}{\varphi(2\gamma+1)}$, as we may, if γ is large enough, (since $\sigma_0-1\geqslant\dfrac{1}{e^{\varphi(2\gamma+1)}}$, where $\varphi(t)\to\infty$ as $t\to\infty$, while $\dfrac{\varphi(t)}{\theta(t)}=o(e^{\varphi(t)})$), the numerator in (72) is positive, and it follows that

$$1-\beta\geqslant\frac{\theta(2\gamma+1)}{1240A_3\varphi(2\gamma+1)}$$

as claimed, on the assumption that $\beta>\sigma_0-\frac{1}{2}r$, which was used in (70). If $\beta\leqslant\sigma_0-\frac{1}{2}r$, then the choice of σ_0 gives us

$$\beta\leqslant\sigma_0-\frac{1}{2}r=1+\frac{1}{40A_3}\frac{\theta(2\gamma+1)}{\varphi(2\gamma+1)}-\frac{1}{2}\theta(2\gamma+1),$$

which leads again to (68). Hence the lemma is proved.

With the help of Theorem 3, and Landau's Lemmas 2 and 3, we are now in a position to prove

THEOREM 4 (LITTLEWOOD). *There exists a positive constant A, such that*

$$\zeta(s)\neq 0, \quad \text{for} \quad \sigma>1-\frac{A\log\log t}{\log t}, \qquad t>t_0. \tag{73}$$

PROOF. By (50), we may choose, in Lemma 3,

$$\theta(t)=\frac{(\log\log t)^2}{\log t}, \qquad \varphi(t)=A\log\log t, \qquad A>1, \qquad t>t_0,$$

and the theorem follows.

REMARK. Because of (15) in Ch. II, we can choose $\theta(t)=\frac{1}{2}$, $\varphi(t)=\log(t+2)$ in Lemma 3, and obtain the following result, which is weaker than Theorem 4:

COROLLARY. *There exists a constant A such that $\zeta(s)$ is not zero for*

$$\sigma\geqslant 1-\frac{A}{\log t}, \qquad t>t_0.$$

NOTE. The same letter A may denote different constants at different occurrences.

Since $\zeta(1+it)\neq 0$ for $0\leqslant t\leqslant t_0$ (see Notes on § 1), (73) implies that $\zeta(s)\neq 0$ in the region (1).

§ 5. Applications of Littlewood's theorem. We shall use Littlewood's theorem to obtain an estimate of $\psi_1(x)-\frac{1}{2}x^2$, where

$$\psi_1(x)=\int_0^x \psi(u)\,du, \qquad x>1,$$

ψ being Chebyshev's function. We shall derive from it an estimate for $\psi(x)-x$. A first step in this derivation is the following

THEOREM 5 (LANDAU). *Suppose that η is a decreasing function of t, for $t\geqslant 0$, and has a derivative η', which is continuous for $t\geqslant 0$. Suppose that*

$$0<\eta(t)\leqslant\tfrac{1}{2}; \tag{74}$$

$$\eta'(t)\to 0, \quad \textit{as} \quad t\to\infty; \tag{75}$$

$$\frac{1}{\eta(t)}=O(\log t), \quad \textit{as} \quad t\to\infty. \tag{76}$$

Suppose that $\zeta(s)$ has no zero in the domain

$$\sigma>1-\eta(|t|).$$

Let α be a fixed number, such that $0<\alpha<1$. Then

$$\frac{\zeta'(s)}{\zeta(s)}=O(\log^2|t|), \tag{77}$$

uniformly in the domain $\sigma\geqslant 1-\alpha\eta(|t|)$, as $t\to\infty$.

PROOF. We may assume that $t>0$, and confine attention to the range $1-\alpha\eta(t)\leqslant\sigma\leqslant 1+\alpha\eta(t)$ of σ, for if $\sigma\geqslant 1+\alpha\eta(t)$, then

$$\left|\frac{\zeta'(s)}{\zeta(s)}\right| \leqslant \sum_{n=1}^{\infty} \frac{\Lambda(n)}{n^{1+\alpha\eta}} = -\frac{\zeta'}{\zeta}(1+\alpha\eta) < \frac{c_1}{\alpha\eta},$$

where $\eta=\eta(t)$, because $\frac{\zeta'}{\zeta}$ has a simple pole at $s=1$. From (76) it follows that

$$\frac{\zeta'(s)}{\zeta(s)} = O(\log t), \quad \text{as} \quad t\to\infty .$$

Since $\zeta(s)$ is regular, and different from zero, in the simply connected domain D defined by $t>0$, $\sigma>1-\eta(t)$, the function $Z(s)=\log\zeta(s)$ is regular in D, and

$$Z(s)=\log\zeta(s)=\sum_{p,m}\frac{1}{mp^{ms}}, \qquad \sigma>1,$$

where m runs through all positive integers, and p through all primes. Our object is to estimate $Z'(s)$. We shall do this by utilizing an estimate for $\operatorname{Re} Z(s)$, and appealing to the lemma of Borel-Carathéodory (Ch. II, Lemma B).

Let $T>1$, and $\eta(T)=H$. We choose the point $s_0=1+\alpha H+iT$, $0<\alpha<1$, as the centre of two concentric circles of radii r and R, where $r=2\alpha H$, and $R=\frac{1}{2}(1+3\alpha)H$, so that $r<R$. The smaller circle then intersects the line $t=T$, at the points $1-\alpha H+iT$ and $1+3\alpha H+iT$, while the larger circle intersects at the points $1-\frac{1}{2}(1+\alpha)H+iT$, and $1+\frac{1}{2}(1+5\alpha)H+iT$. If T is large enough, both these circles lie in D. Since $R<2H\leqslant 1$ by (74), $T>1$, and η is a decreasing function, it will be sufficient to show that the point

$$s'=\sigma'+it'=1-\tfrac{1}{2}(1+\alpha)H+i(T+R)$$

lies in D, or that the condition $\sigma'>1-\eta(t')$ is satisfied for large T. But this is so, because

$$\begin{aligned}\sigma'-1+\eta(t') &= -\tfrac{1}{2}(1+\alpha)H+\eta(T+R)\\ &= -\tfrac{1}{2}(1+\alpha)H+\eta(T)+R\eta'(\tau), \qquad T<\tau<T+R,\\ &= \left(\tfrac{1}{2}(1-\alpha)+\tfrac{1}{2}(1+3\alpha)\eta'(\tau)\right)H\\ &> 0,\end{aligned}$$

for all $T>T_1$, by (75). Hence, if $T>T_1$, then $Z(s)$ is regular for $|s-s_0|<R$.

Throughout the disc $|s-s_0|<R$, we have $\sigma>\frac{1}{2}$, and $T-1<t<T+1$. Hence, by Theorem 5, Ch. II, we have

$$\operatorname{Re} Z(s)=\log|\zeta(s)|<\log(ct^{\frac{1}{2}})<\log T, \tag{78}$$

for $T > T_2$, throughout the disc $|s - s_0| < R$. Further

$$|\mathrm{Re}\, Z(s_0)| \leqslant |Z(s_0)| \leqslant \sum_{p,m} \frac{1}{m p^{m(1+\alpha H)}}$$

$$< \sum_{n=2}^{\infty} \frac{1}{n^{1+\alpha H}} < \frac{1}{\alpha H}.$$

By an appeal to the lemma of Borel-Carathéodory (Ch. II), we deduce that

$$|Z'(s)| < \frac{2R}{(R-r)^2} (\log T - \mathrm{Re}\, Z(s_0)) < \frac{4(1+3\alpha)}{(1-\alpha)^2 H} \left(\log T + \frac{1}{\alpha H} \right),$$

throughout $|s - s_0| \leqslant r$. This holds, in particular, on the radius extending from the point s_0 to the point $1 - \alpha H + iT$; that is, for $s = \sigma + iT$, $1 - \alpha \eta(T) \leqslant \sigma \leqslant 1 + \alpha \eta(T)$. Since $\frac{1}{H} = O(\log t)$ by (76), we obtain (77).

The estimate of $\frac{\zeta'}{\zeta}$ obtained in Theorem 5 is used to prove the following

THEOREM 6. *Under the conditions of Theorem 5, we have*

$$\psi_1(x) - \tfrac{1}{2} x^2 = O(x^2 e^{-\alpha \omega(x)}), \quad \text{as} \quad x \to \infty, \tag{79}$$

where $\omega(x)$ *is the minimum of* $\eta(t) \log x + \log t$ *for* $t \geqslant 1$.

PROOF (INGHAM). The minimum $\omega(x)$ exists for every $x > 0$, since $\eta(t) \log x + \log t$ is continuous for $t \geqslant 1$, and tends to infinity with t.

If $x > 1$, we have

$$\psi_1(x) = \int_0^x \psi(u)\, du = \int_1^x \psi(u)\, du = \sum_{n \leqslant x} \Lambda(n)(x - n), \tag{80}$$

and from the classical formula of Perron, we have

$$\psi_1(x) = \frac{1}{2\pi i} \int_{c - i\infty}^{c + i\infty} \frac{x^{s+1}}{s(s+1)} \left(- \frac{\zeta'(s)}{\zeta(s)} \right) ds, \quad c > 1. \tag{81}$$

If $\mathscr{C}$ denotes the curve defined by $\sigma = 1 - \alpha \eta(|t|)$, $0 < \alpha < 1$, then

$$\psi_1(x) = \frac{x^2}{2} - \frac{1}{2\pi i} \int_{\mathscr{C}} \frac{\zeta'(s)}{\zeta(s)} \cdot \frac{x^{s+1}}{s(s+1)}\, ds. \tag{82}$$

This results from (81) by an application of Cauchy's theorem, which is possible because the integrand is regular in the region bounded by $\mathscr{C}$

and the line $\sigma=c$, except for $s=1$, and, by Theorem 5, it is uniformly $O(t^{-2}\log^2|t|)$, as $t\to\pm\infty$, for fixed x. We write

$$\psi_1(x)=\tfrac{1}{2}x^2-J, \tag{83}$$

and seek to estimate J. Because of symmetry, we may confine attention to that part of $\mathscr{C}$ which lies in the upper half-plane. On that part of $\mathscr{C}$, we have, by Theorem 5,

$$\left|\frac{\zeta'(s)}{\zeta(s)}\right|<c_1\log^2(t+2),$$

and

$$\left|\frac{ds}{dt}\right|=|-\alpha\eta'(t)+i|<c_2,$$

because of (75). Hence

$$|J|<c_3\int_0^\infty x^2\cdot\frac{x^{-\alpha\eta(t)}\log^2(t+2)}{(t+1)^2}\,dt$$

$$=c_3\int_1^\infty x^2\cdot\frac{x^{-\alpha\eta(u-1)}\log^2(u+1)}{u^2}\,du.$$

Since $\eta(u-1)\geqslant\eta(u)$, and $x>1$, it follows from (83) that

$$\left|\frac{\psi_1(x)}{x^2}-\frac{1}{2}\right|<c_3\int_1^\infty e^{-\alpha\eta(u)\log x-\alpha\log u}\frac{\log^2(u+1)}{u^{2-\alpha}}\,du$$

$$\leqslant c_3\,e^{-\alpha\omega(x)}\int_1^\infty\frac{\log^2(u+1)}{u^{2-\alpha}}\,du,\qquad 2-\alpha>1$$

$$=c_4\cdot e^{-\alpha\omega(x)},$$

which proves the theorem.

The transition from ψ_1 to ψ is effected by a general analytical argument.

THEOREM 7. *Under the conditions of Theorem 5, we have*

$$\psi(x)-x=O(x\,e^{-\frac{1}{2}\alpha\omega(x)}), \tag{84}$$

and

$$\pi(x)-\mathrm{li}\,x=O(x\,e^{-\frac{1}{2}\alpha\omega(x)}). \tag{85}$$

PROOF (INGHAM). We note that $\omega(x)$ is a strictly increasing function of x, for $x>0$, and so is the function $\log x-\omega(x)$. For let $x_1>x_2>0$, and let t_1, t_2 be values of t for which the minima $\omega(x_1), \omega(x_2)$ are attained. Then

$$\omega(x_2) \leqslant \eta(t_1)\log x_2+\log t_1 <\eta(t_1)\log x_1+\log t_1=\omega(x_1),$$

while

$$\begin{aligned}\omega(x_1) &\leqslant \eta(t_2)\log x_1+\log t_2=\omega(x_2)+\eta(t_2)(\log x_1-\log x_2)\\ &<\omega(x_2)+(\log x_1-\log x_2).\end{aligned}$$

Since $\omega(1)=0$, and $\log 1-\omega(1)=0$, we deduce that

$$0<\omega(x)<\log x, \quad \text{for} \quad x>1. \tag{86}$$

Now suppose that $x>2$. Let h be a function of x, such that $0<h<\frac{1}{2}x$. Since ψ is an increasing function,

$$\frac{1}{h}\int_{x-h}^{x}\psi(u)\,du \leqslant \psi(x) \leqslant \frac{1}{h}\int_{x}^{x+h}\psi(u)\,du. \tag{87}$$

By definition of ψ_1 (see (80)), the two integrals are

$$\frac{\psi_1(x\pm h)-\psi_1(x)}{\pm h}=x\pm\tfrac{1}{2}h+O\left(x^2\frac{e^{-\alpha\omega(\frac{1}{2}x)}}{h}\right), \tag{88}$$

by Theorem 6, since $\frac{1}{2}x<x-h<x<x+h<\dfrac{3x}{2}$, and $\omega(x)$ is an increasing function of x.

In the O-term in (88) we may replace $\omega(\frac{1}{2}x)$ by $\omega(x)$, for $\log u-\omega(u)$ is an increasing function of u, and

$$\begin{aligned}e^{-\alpha\omega(\frac{1}{2}x)}&=(\tfrac{1}{2}x)^{-\alpha}e^{\alpha(\log\frac{1}{2}x-\omega(\frac{1}{2}x))}\\ &\leqslant(\tfrac{1}{2}x)^{-\alpha}e^{\alpha(\log x-\omega(x))}\\ &=2^{\alpha}e^{-\alpha\omega(x)}.\end{aligned}$$

Thus from (88) and (87), we have

$$\psi(x)=x+O(h)+O(h^{-1}x^2e^{-\alpha\omega(x)}).$$

Taking $h=\frac{1}{2}xe^{-\frac{1}{2}\alpha\omega(x)}$, as we may, we get (84).

To prove (85), we define the function $\Pi(x)$, for $x>0$, by the relation

$$\Pi(x)=\sum_{p^m\leqslant x}\frac{1}{m}=\pi(x)+\tfrac{1}{2}\pi(x^{\frac{1}{2}})+\tfrac{1}{3}\pi(x^{\frac{1}{3}})+\cdots. \tag{89}$$

Since

$$\log\zeta(s) = \sum_{p,m} \frac{1}{mp^{ms}}, \qquad \sigma > 1,$$

we have, by Abel's partial summation formula, (Notes on Ch. I),

$$\log\zeta(s) = s\int_1^\infty \frac{\Pi(x)}{x^{s+1}}\,ds, \qquad \sigma > 1.$$

Further

$$\Pi(x) = \sum_{2\leqslant n\leqslant x} \frac{\Lambda(n)}{\log n} = \frac{\psi(x)}{\log x} + \int_2^x \frac{\psi(u)\,du}{u\log^2 u},$$

and, if $x > 2$,

$$\int_2^x \frac{du}{\log u} = \frac{x}{\log x} - \frac{2}{\log 2} + \int_2^x \frac{u\,du}{u\log^2 u}.$$

Hence

$$\Pi(x) - \operatorname{li} x = \frac{\psi(x) - x}{\log x} + O(1) + \int_2^x \frac{\psi(u) - u}{u\log^2 u}\,du,$$

where $\operatorname{li} x$ is defined by (6). By (84), which we have already proved, we deduce that

$$\Pi(x) - \operatorname{li} x = O(xe^{-\frac{1}{2}\alpha\omega(x)}) + O(1) + O\left(\int_2^x e^{-\frac{1}{2}\alpha\omega(u)}\,du\right),$$

and the term $O(1)$ may be omitted, since $0 < \omega(x) < \log x$, and

$$xe^{-\frac{1}{2}\alpha\omega(x)} > xe^{-\frac{1}{2}\alpha\log x} = x^{1-\frac{1}{2}\alpha} > x^{\frac{1}{2}} > 1, \tag{90}$$

since we have supposed that $x > 2$.

Further we have

$$\int_2^x e^{\frac{1}{2}\alpha(\log u - \omega(u))}\frac{du}{u^{\frac{1}{2}\alpha}} \leqslant e^{\frac{1}{2}\alpha(\log x - \omega(x))}\int_2^x \frac{du}{u^{\frac{1}{2}\alpha}} = O(xe^{-\frac{1}{2}\alpha\omega(x)}),$$

since $1-\frac{1}{2}\alpha>0$. Hence

$$\Pi(x)-\operatorname{li}x=O(x e^{-\frac{1}{2}\alpha\omega(x)}), \tag{91}$$

and since (89) implies that

$$\Pi(x)-\pi(x) = \sum_{m=2}^{M} \frac{\pi(x^{1/m})}{m}, \qquad M = \left[\frac{\log x}{\log 2}\right],$$
$$= O(x^{\frac{1}{2}})+O(M x^{\frac{1}{3}})=O(x^{\frac{1}{2}}), \tag{92}$$

we have (85) from (91), (92) and (90).

Littlewood's theorem permits us to prove the following

THEOREM 8. *As* $x\to\infty$, *we have*

$$\psi(x)-x = O(x e^{-a\sqrt{\log x \log\log x}}), \tag{93}$$

and

$$\pi(x)-\operatorname{li}x=O(x e^{-a\sqrt{\log x \log\log x}}), \tag{94}$$

where a is a positive, absolute constant.

PROOF. Because of Littlewood's theorem, we can choose the function η of Theorems 5, 6, 7, as follows:

$$\eta(t) = \begin{cases} a_2\cdot\dfrac{\log\log t}{\log t}, & t\geqslant e^e, \quad a_2 \text{ a constant}, \\ a_2/e, & 0\leqslant t\leqslant e^e. \end{cases}$$

By suitable choice of a_2, we can verify that η satisfies the conditions postulated in Theorem 5. If we set $\log\xi=(\log x)^{\frac{1}{2}}$, and choose x so large that $\xi\geqslant e^e$, then, since $(A+B)^2\geqslant 4AB$,

$$\eta(t)\log x+\log t \geqslant \begin{cases} 2(a_2\log x \log\log t)^{\frac{1}{2}}\geqslant(2a_2\log x\log\log x)^{\frac{1}{2}} & \text{for } t\geqslant\xi, \\ \eta(\xi)\log x=\frac{1}{2}a_2(\log x)^{\frac{1}{2}}\log\log x & \text{for } 1\leqslant t\leqslant\xi. \end{cases}$$

Hence $\omega(x)\geqslant(2a_2\log x\log\log x)^{\frac{1}{2}}$ for all sufficiently large x. Theorem 8 is now a consequence of Theorem 7.

Notes on Chapter III

§ 1. The non-vanishing of $\zeta(s)$ on the line $\sigma=1$ is sufficient to prove the prime number theorem (of which it is an easy consequence) by the Wiener-Ikehara method, see the author's *Introduction*, Ch. XI. The arrangement of the proof given there that $\zeta(1+it)\neq 0$ for $t\neq 0$ is due to F. Mertens, *Sitzungsberichte Akad. Wiss. Wien*, Math.-naturwiss. Klasse, Bd. 107, Abt. 2a (1898) 1429—1434.

For a proof of the existence of the limit in (6), see, for instance, Landau's *Primzahlen*, I, 27. For the step following (7), see ibid., 85.

The fact that $\zeta(s) \neq 0$ in the domain $\sigma > 1 - \frac{a}{\log(|t|+2)}$, where a is a positive, absolute constant, was first proved by de la Vallée-Poussin in 1899, see *Mémoires couronnés et autres mémoires publiés par l'Académie royale des Sciences, des Lettres et des Beaux-Arts de Belgique*, Bd. 59, No. 1, 74. He also proved the consequent estimate

$$\pi(x) - \operatorname{li} x = O(x e^{-\alpha\sqrt{\log x}}).$$

See the Corollary at the end of § 4, as well as Theorems 19 and 23 of Ingham's *Tract*.

P. Turán has proved a theorem in the opposite direction, and deduced the existence of a zero-free region from an estimate of the error term, *Acta Math. Acad. Sci. Hungar.* 1 (1950), 155—166.

That the Riemann hypothesis implies (12) and (13) was first proved by H. von Koch, *Acta Math.* 24 (1901), 159—182. See also Theorem 30 of Ingham's *Tract*.

§ 2—3. Weyl's method originated in his two papers, *Math. Annalen*, 77 (1916), 313—52; *Math. Zeitschrift*, 10 (1921), 88—101; *Gesammelte Abh.* I, 563; II, 181. See also G. H. Hardy and J. E. Littlewood, *Proc. International Congress of Math.*, Cambridge (1912), I, 223—9. A comprehensive formulation is due to Landau, *Vorlesungen*, II, 31—46; I, Satz 265.

In the proof of Theorem 1 by induction, we say that if (19) is true with $k-1$ in place of k, then it can be applied to estimate the sum S_1 in (20). To see this one has only to put $l = m - \max(r_1+1, 1) + 1$, so that l runs from 1 to μ', where $\mu' = \min(r_1+\mu, \mu) - \max(r_1+1, 1) + 1 \leqslant \mu - 1 + 1 = \mu$, while the coefficient of the highest power of l in the exponent remains unchanged.

In the proof of (54) we use the following proposition. Let $b_1 \geqslant b_2 \geqslant \cdots \geqslant b_n \geqslant \cdots \geqslant 0$, and $s_n = \sum_{M<m\leqslant n} a_m$, where the a's are complex numbers, M is a non-negative integer, and $s_M = 0$. We then have

$$\left|\sum_{n=M+1}^{N} a_n b_n\right| \leqslant b_M \max_{M<n\leqslant N} \left|\sum_{m=M+1}^{n} a_m\right|,$$

for

$$\sum_{n=M+1}^{N} a_n b_n = \sum_{M<n\leqslant N} (s_n - s_{n-1}) b_n = s_N b_N + \sum_{M<n\leqslant N-1} s_n(b_n - b_{n+1}),$$

so that

$$\left|\sum_{n=M+1}^{N} a_n b_n\right| \leqslant |s_N| b_N + \sum_{n=M+1}^{N-1} |s_n|(b_n - b_{n+1})$$

$$\leqslant \left(b_N + \sum_{n=M+1}^{N-1} (b_n - b_{n+1})\right) \max_{M<n\leqslant N} |s_n| = b_{M+1} \max_{M<n\leqslant N} |s_n|$$

$$\leqslant b_M \max_{M<n\leqslant N} |s_n|.$$

Similarly, if $0\leqslant b_1 \leqslant b_2 \leqslant \cdots \leqslant b_n \leqslant \cdots$, then

$$\left|\sum_{n=M+1}^{N} a_n b_n\right| \leqslant 2 b_N \max_{M<n\leqslant N} \left|\sum_{m=M+1}^{n} a_m\right|,$$

for $b_{n+1} - b_n \geqslant 0$, and $\left|\sum_{n=M+1}^{N} a_n b_n\right| \leqslant |s_N| b_N + \sum_{M<n\leqslant N-1} |s_n|(b_{n+1} - b_n)$ $\leqslant 2 b_N \max_{M<n\leqslant N} |s_n|$.

Taking $a_n = e^{-it\log n}$, $t>0$, $b_n = n^{-\sigma}$, $\sigma\geqslant 0$, and a, b positive integers such that $b-a\geqslant 1$, we get

$$\sum_{n=a+1}^{b} n^{-s} = O\left(a^{-\sigma} \max_{a<c\leqslant b} \left|\sum_{n=a+1}^{c} e^{-it\log n}\right|\right).$$

This is used in the proof of (54).

The constant $\frac{63}{64}$ in Theorem 3 is *not* the best possible.

§ 4. Lemma 2 is due to Landau, *Vorlesungen*, II, Satz 374. Lemma 3 is given in this form by Titchmarsh, *Zeta-function*, 50, Theorem 3.10, though the general idea derives from Landau's earlier formulation, *Math. Zeitschrift*, 20 (1924), 100. This proof of Littlewood's theorem which makes no use of a knowledge of $\zeta(s)$ for $\sigma<0$ is due to Landau, *Vorlesungen*, II, Kap. 2. The original statement by Littlewood is in *Proc. London Math. Soc.* (2) 20 (1922), xxii—xxviii. (Records, Feb. 10, 1921). See Ingham's remarks in his *Tract*, 66.

§ 5. Theorems 5, 6, 7, in this form, are given by Ingham in his *Tract*, 60—63. His proofs are influenced, however, by Landau's earlier formulation, *Vorlesungen*, II, Satz 376; *Math. Zeitschrift*, 20 (1924), 123. Theorem 8 was extended by Landau to obtain an estimate of the error term in the prime number theorem for arithmetical progressions. His analysis applies not only to $\zeta(s)$ but to Dirichlet's L-functions. Thus Satz 403 of his *Vorlesungen*, II, proves that if $(k,l)=1$, then

$$\pi(x;k,l) = \sum_{\substack{p\leqslant x \\ p\equiv l(\mathrm{mod}\,k)}} 1 = \frac{\mathrm{li}\,x}{\varphi(k)} + O(x e^{-c\sqrt{\log x \log\log x}}),$$

where the constant involved in the O depends on k. This is a refinement of de la Vallée-Poussin's earlier result that

$$\pi(x;k,l)=\frac{\operatorname{li}x}{\varphi(k)}+O(xe^{-c\sqrt{\log x}}).$$

Since $\pi(x;1,1)=\pi(x)$, which is defined in Chapter I, this gives as a special case

$$\pi(x)-\operatorname{li}x=O(xe^{-c\sqrt{\log x}})=O\left(\frac{x}{\log^{\Delta}x}\right)$$

for any fixed positive Δ. For an elementary proof of this estimate, more in tune with the method of Chapter I, see E. Bombieri, *Riv. Mat. Univ. Parma* (2) 3 (1962); *Istituto Lombardo (Rend.-Sci.)* A 96 (1962), 343–350; and E. Wirsing, *Journal für Math.* 214/215 (1963), 1—18.

Formula (81) is a special case of a formula for the generalized partial sums of the coefficients of Dirichlet series, referred to as *Perron's formula*. See, for instance, G. H. Hardy and M. Riesz, *The general theory of Dirichlet series*, Cambridge Tract, 18 (1915), Theorems 13, 24, 29, 39, 40; and K. Chandrasekharan and S. Minakshisundaram, *Typical means*, Oxford (1952), 81. An explicit proof of (81) is given, by contour integration, in Ingham's *Tract*, 31.

Chapter IV

Vinogradov's method

§ 1. A refinement of Littlewood's theorem. Weyl's method of estimating trigonometric sums was used in the previous chapter to prove Littlewood's theorem on the zero-free region of $\zeta(s)$. Littlewood's theorem was used, in turn, to obtain the following estimate of the error term in the prime number theorem:

$$\pi(x)-\operatorname{li} x=O\left(x e^{-a\sqrt{\log x \log\log x}}\right),$$

for a positive, absolute constant a. A powerful refinement of Weyl's method was effected by I. M. Vinogradov, who applied it to the solution of a variety of problems in number theory. We shall describe the essentials of that method in this chapter, and use it to deduce Chudakov's refinement of Littlewood's theorem, to the effect that there exists a constant $A_1>0$, such that

$$\zeta(s)\neq 0, \quad \text{for} \quad \sigma\geqslant 1-\frac{A_1}{\log^{\frac{3}{4}} t(\log\log t)^{\frac{3}{4}}}, \qquad t\geqslant t_1.$$

This result can again be used, as in the previous chapter, to prove that

$$\pi(x)-\operatorname{li} x=O\left(x e^{-c(\log x)^{\frac{4}{7}}(\log\log x)^{-\frac{3}{7}}}\right),$$

for a constant $c>0$. This is an improvement on the approximation of $\pi(x)$ obtained in the previous chapter.

§ 2. An outline of the method. Let n be an integer, and

$$f(n)=\alpha_k n^k+\cdots+\alpha_1 n+\alpha_0$$

be a polynomial in n, of degree $k\geqslant 2$, with real coefficients. Let

$$S(q)=\sum_{a<n\leqslant a+q} e^{2\pi i f(n)}, \tag{1}$$

where a and q are integers, and

$$J(q,l)=\int_0^1\cdots\int_0^1|S(q)|^{2l}\,d\alpha_1\ldots d\alpha_k, \tag{2}$$

where l is a positive integer.

Trivially we have

$$|S(q)|\leqslant q, \qquad J(q,l)\leqslant q^{2l}. \tag{3}$$

Less trivially we can prove that

$$J(q,l)\leqslant k!\,q^{2l-k}, \qquad l\geqslant k. \tag{4}$$

Since

$$\{S(q)\}^l=\sum_{n_1,\ldots,n_l} e^{2\pi i\alpha_k(n_1^k+\cdots+n_l^k)+\cdots},$$

and

$$|S(q)|^{2l}=\sum_{m_1,\ldots,n_l} e^{2\pi i\alpha_k(m_1^k+\cdots+m_l^k-n_1^k-\cdots-n_l^k)+\cdots}, \tag{5}$$

and since, for an integer h,

$$\int_0^1 e^{2\pi ihx}\,dx=\begin{cases}1, & \text{if } h=0,\\ 0, & \text{if } h\neq 0,\end{cases}$$

it follows that $J(q,l)$ is equal to the number of solutions in integers of the system of equations

$$m_1^h+\cdots+m_l^h=n_1^h+\cdots+n_l^h, \qquad h=1,2,\ldots,k, \tag{6}$$

where $a<m_\nu\leqslant a+q$, $a<n_\nu\leqslant a+q$. In particular, $J(q,k)$ is equal to the number of solutions of the system

$$m_1^h+\cdots+m_k^h=n_1^h+\cdots+n_k^h, \qquad h=1,2,\ldots,k.$$

These equations imply that the numbers (n_ν) are all equal, in some order, to the numbers (m_ν), so that only the set (m_ν), $\nu=1,2,\ldots,k$ can be chosen arbitrarily. Therefore $J(q,k)\leqslant k!\,q^k$. If this is combined with the inequality $J(q,l)\leqslant q^{2(l-k)}J(q,k)$, which results from (2), we get (4).

Clearly $J(q,l)$ is a non-decreasing function of q, and it is independent of a, since, if we put $M_\nu=m_\nu-a$, and $N_\nu=n_\nu-a$, in (6), we obtain

$$\sum_{\nu=1}^{l}(M_\nu+a)^h=\sum_{\nu=1}^{l}(N_\nu+a)^h, \qquad h=1,2,\ldots,k,$$

and, on expanding the hth powers, we find that this system is equivalent to

$$\sum_{\nu=1}^{l} M_\nu^h = \sum_{\nu=1}^{l} N_\nu^h, \qquad h=1,2,\ldots,k,$$

where $0<M_\nu\leqslant q$, $0<N_\nu\leqslant q$.

The first step in Vinogradov's method is to obtain an estimate for $J(q,l)$, defined in (2), which is stronger than (4). This is done by finding an upper bound for the number of solutions of the system of equations in (6), and more simply, after Linnik, Karacuba and Korobov, of a system of congruences. The result is referred to as *Vinogradov's mean-value theorem* (Theorem 2).

The second step is to use the mean-value theorem to estimate sums of the form

$$\sum_{a<n\leqslant a+Q} e^{2\pi i F(n)},$$

where F is a real-valued function of a real variable, which is continuously differentiable a certain number of times, with its derivatives suitably restricted in size in the range considered, and Q likewise restricted. This is achieved by approximating to F by a suitable polynomial f.

By choosing $F(x) = -\dfrac{t\log x}{2\pi}$, $t>0$, one applies this estimate to sums of the form

$$\sum_{a<n\leqslant b} n^{-it},$$

and therefore to sections of the zeta-function in the critical strip, namely

$$\sum_{a<n\leqslant b} n^{-\sigma-it}, \qquad 0<\sigma<1, \qquad t>0,$$

so as to obtain a result of the following type:

$$\zeta(s)\neq 0, \quad \text{for } \sigma>1-\frac{1}{(\log t)^a}, \qquad t>t_1,$$

for a certain constant $a<1$, and for $t_1>2$. This is obviously stronger than Littlewood's theorem.

§ 3. Vinogradov's mean-value theorem. We shall first obtain an inequality for $J(q,l)$ in terms of $J(q_1,l_1)$, where $q_1<q$ and $l_1<l$. We require a preliminary

LEMMA 1. *Let T be the number of solutions, in integers, of the system of congruences*

$$\left.\begin{array}{l} x_1+\cdots+x_k\equiv\mu_1(\operatorname{mod} p)\\ \cdots\cdots\cdots\cdots\cdots\cdots\cdots\\ x_1^k+\cdots+x_k^k\equiv\mu_k(\operatorname{mod} p^k)\end{array}\right\}, \tag{7}$$

where $0 \leqslant x_j \leqslant Mp^k - 1$, *for* $j = 1, 2, \ldots, k$; M *an integer*, $M \geqslant 1$; $\mu_1, \mu_2, \ldots, \mu_k$ *given integers*; p *a prime*, $p > k$, *and* $x_j \not\equiv x_{j'} (\bmod p)$ *for* $j \neq j'$. *Then*

$$T \leqslant k!\, M^k p^{\frac{1}{2}k(k-1)}. \tag{8}$$

Proof. If we write the integers x_j in the form

$$x_j = x_{j1} + p x_{j2} + \cdots + p^{k-1} x_{jk} + p^k x_{j,k+1}, \qquad j = 1, 2, \ldots, k,$$

where $0 \leqslant x_{j\nu} \leqslant p - 1$, $\nu = 1, 2, \ldots, k$, and $0 \leqslant x_{j,k+1} \leqslant M - 1$, then we have $x_{j1} \neq x_{j'1}$ for $j \neq j'$, and the integers $x_{11}, \ldots, x_{k1}$ satisfy the system of congruences

$$\left.\begin{array}{l} x_{11} + \cdots + x_{k1} \equiv \mu_1 \\ \cdots\cdots\cdots\cdots\cdots\cdots \\ x_{11}^k + \cdots + x_{k1}^k \equiv \mu_k \end{array}\right\} (\bmod p). \tag{9}$$

If T_1 denotes the number of solutions of this system, then $T_1 \leqslant k!$, since the system has a unique solution if any, except for the order of the indeterminates $x_{11}, \ldots, x_{k1}$, all of which are supposed to be distinct. Keeping one such solution fixed, we consider the system of congruences $\bmod p^2$, namely

$$\left.\begin{array}{l} (x_{11} + p x_{12})^2 + \cdots + (x_{k1} + p x_{k2})^2 \equiv \mu_2 \\ \cdots\cdots\cdots\cdots\cdots\cdots\cdots\cdots\cdots\cdots \\ (x_{11} + p x_{12})^k + \cdots + (x_{k1} + p x_{k2})^k \equiv \mu_k \end{array}\right\} (\bmod p^2). \tag{10}$$

This reduces to a system of $(k-1)$ linear congruences in $x_{12}, \ldots, x_{k2}$, namely

$$\left.\begin{array}{l} x_{11} x_{12} + \cdots + x_{k1} x_{k2} \equiv \mu_2' \\ \cdots\cdots\cdots\cdots\cdots\cdots\cdots \\ x_{11}^{k-1} x_{12} + \cdots + x_{k1}^{k-1} x_{k2} \equiv \mu_k' \end{array}\right\} (\bmod p). \tag{11}$$

Since $x_{11}, \ldots, x_{k1}$ are distinct, at least $k-1$ of them are different from zero. We may assume, for instance, that $x_{11}, \ldots, x_{k-1,1}$ are non-zero. Then

$$\begin{vmatrix} x_{11} & \ldots x_{k-1,1} \\ \cdots & \cdots \\ x_{11}^{k-1} & \ldots x_{k-1,1}^{k-1} \end{vmatrix} = x_{11} \ldots x_{k-1,1} \prod_{1 \leqslant j < j' \leqslant k-1} (x_{j1} - x_{j'1}) \not\equiv 0 (\bmod p).$$

Therefore, for any fixed x_{k2}, the integers $x_{12}, \ldots, x_{k-1,2}$ are uniquely determined from the system (11). If T_2 denotes the number of solutions of this system, then $T_2 = p$.

Similarly we consider congruences $\bmod p^3$, and get a system of $k-2$ congruences in $x_{13}, \ldots, x_{k3}$, with the number of solutions $T_3 = p^2$.

In the same way, we get $T_4 = p^3, \ldots, T_k = p^{k-1}$, while $T_{k+1} = M^k$. Since $T \leqslant T_1 \ldots T_{k+1}$, it follows that

$$T \leqslant k!\, p^{\frac{k(k-1)}{2}} M^k,$$

as claimed in (8).

We use Lemma 1 to prove an inequality between $J(q,l)$ and $J(q_1,l_1)$, for $q_1 < q$, $l_1 < l$.

THEOREM 1 (KARACUBA). *Let k, l, and q be positive integers, such that*

$$k \geqslant 2, \quad l \geqslant k^2 + k, \quad q \geqslant 2^{-k}(2k)^{2k} = (2k^2)^k. \tag{12}$$

Then

$$J(q,l) < 3\, l^{2k} q_1^{2k} p_1^{2l - \frac{1}{2}k(k+1)} J(q_1, l-k), \tag{13}$$

where p_1 is any prime, such that

$$\tfrac{1}{2} q^{1/k} \leqslant p_1 \leqslant q^{1/k}, \tag{14}$$

and

$$q_1 = [q p_1^{-1}] + 1. \tag{15}$$

PROOF. The existence of p_1 in the interval indicated in (14) is ensured by *Bertrand's postulate*, namely that for every $x \geqslant 2$, there exists a prime p, such that $x < p \leqslant 2x$.

By definition we have $q \leqslant p_1 q_1$, so that $J(q,l) \leqslant J(p_1 q_1, l)$. But $q \neq p_1 q_1$, hence

$$J(q,l) < J(p_1 q_1, l), \tag{16}$$

where $J(p_1 q_1, l)$ is the number of solutions, in integers, of the system of equations

$$M_1^h + M_2^h + \cdots + M_l^h = N_1^h + N_2^h + \cdots + N_l^h,$$

for $h = 1, 2, \ldots, k$, and $0 \leqslant M_j, N_j \leqslant p_1 q_1 - 1$, for $j = 1, 2, \ldots, l$. Equivalently, $J(p_1 q_1, l)$ is equal to the number of solutions of the system of equations

$$(m_1 + p_1 a_1)^h + \cdots + (m_l + p_1 a_l)^h = (n_1 + p_1 b_1)^h + \cdots + (n_l + p_1 b_l)^h, \tag{17}$$

for $h = 1, 2, \ldots, k$, where a_j and b_j are integers such that $0 \leqslant a_j, b_j \leqslant q_1 - 1$, and m_j, n_j are integers such that $0 \leqslant m_j$, $n_j \leqslant p_1 - 1$. This follows upon writing $M_j = m_j + p_1 a_j$, $N_j = n_j + p_1 b_j$.

We say that a set of integers $(m_1, \ldots, m_l)$ belongs to the first class, if there are at least k distinct integers in it. Otherwise it is said to belong to the second class. Clearly

$$J(p_1 q_1, l) = J_l^{(1)} + J_l^{(2)}, \tag{18}$$

where $J_l^{(1)}$ denotes the number of solutions of (17), such that both $(m_1, \dots, m_l)$ and $(n_1, \dots, n_l)$ belong to the first class, while $J_l^{(2)}$ denotes the number of solutions such that either $(m_1, \dots, m_l)$ or $(n_1, \dots, n_l)$ belongs to the second class.

Let

$$U(m) = \sum_{j=0}^{q_1 - 1} e^{2\pi i g(m + p_1 j)}, \tag{19}$$

where

$$g(n) = \alpha_k n^k + \cdots + \alpha_1 n, \qquad \alpha_k \neq 0.$$

Clearly

$$J_l^{(2)} = \int_0^1 \dots \int_0^1 \left(\sum_{\substack{m_1, \dots, m_l \\ n_1, \dots, n_l}} U(m_1) \dots U(m_l) \overline{U(n_1)} \dots \overline{U(n_l)} \right) d\alpha_1 \dots d\alpha_k, \tag{20}$$

where $m_1, \dots, m_l$, $n_1, \dots, n_l$ vary in such a way that either the set $(m_1, \dots, m_l)$, or the set $(n_1, \dots, n_l)$ belongs to the second class. The number of sets in the second class does not exceed $\frac{1}{2} k^l p_1^{k-1}$, since any set in the second class has at most $k-1$ distinct integers, and $\binom{p_1}{k-1}(k-1)^l < \frac{1}{2} k^l p_1^{k-1}$. Therefore $k^l p_1^{l+k-1}$ is an upper bound for the number of sets $(m_1, \dots, n_l)$ such that either $(m_1, \dots, m_l)$ or $(n_1, \dots, n_l)$ belongs to the second class. Hence, by Hölder's inequality, we have

$$\left| \sum_{m_1, \dots, n_l} U(m_1) \dots \overline{U(n_l)} \right|$$
$$\leqslant \left(\sum_{m_1, \dots, n_l} |U(m_1)|^{2l} \right)^{1/(2l)} \times \cdots \times \left(\sum_{m_1, \dots, n_l} |U(n_l)|^{2l} \right)^{1/(2l)},$$

where trivially we have $|U(\alpha_j)|^{2l} \leqslant \sum_{m=0}^{p_1 - 1} |U(m)|^{2l}$, $\alpha_j = m_j$ or n_j, $1 \leqslant j \leqslant l$, so that

$$\left| \sum_{m_1, \dots, n_l} U(m_1) \dots \overline{U(n_l)} \right| \leqslant k^l p_1^{l+k-1} \sum_{m=0}^{p_1 - 1} |U(m)|^{2l}$$
$$\leqslant k^l q_1^{2k} p_1^{l+k-1} \sum_{m=0}^{p_1 - 1} |U(m)|^{2l - 2k}.$$

This implies, in view of (20), that

$$\begin{aligned} J_l^{(2)} &\leqslant k^l p_1^{l+k-1} q_1^{2k} \int_0^1 \cdots \int_0^1 \sum_{m=0}^{p_1 - 1} |U(m)|^{2l - 2k} d\alpha_1 \dots d\alpha_k \\ &= k^l q_1^{2k} p_1^{l+k} J(q_1, l-k), \end{aligned} \tag{21}$$

if we note that $J(q,l)$ remains unchanged if in (1) and (2) $f(n)$ is replaced by $f(rn)$, where r is an integer different from zero. Since $p_1 \geqslant \frac{1}{2} q^{1/k} \geqslant k^2$, by (14) und (12), and $l \geqslant k^2+k$ by (12), while $k \geqslant 2$, we have $l \geqslant 2k$, so that

$$k^{l-2k} p_1^k \leqslant p_1^{l/2-k} p_1^k = p_1^{l/2} \leqslant p_1^{\,l-\frac{k(k+1)}{2}},$$

which implies that $k^l p_1^{l+k} \leqslant p_1^{\,2l-\frac{k(k+1)}{2}} k^{2k}$. Using this estimate in (21), we get

$$\begin{aligned} J_l^{(2)} &\leqslant k^{2k} q_1^{2k} p_1^{2l-\frac{1}{2}k(k+1)} J(q_1, l-k) \\ &\leqslant l^{2k} q_1^{2k} p_1^{2l-\frac{1}{2}k(k+1)} J(q_1, l-k). \end{aligned} \tag{22}$$

We shall next estimate $J_l^{(1)}$. We denote by $[m_1, \dots, m_k], m_{k+1}, \dots, m_l$ a set with the elements $m_1, \dots, m_k$ distinct, and $m_{k+1}, \dots, m_l$ arbitrary. Clearly every set of the first class is included among the permutations of sets of the form $[m_1, \dots, m_k], m_{k+1}, \dots, m_l$. Hence the number $J_l^{(1)}$ does not exceed the number of solutions of (17) with variables of the form $[m_1, \dots, m_k], m_{k+1}, \dots, m_l$ and $[n_1, \dots, n_k], n_{k+1}, \dots, n_l$, multiplied by $\binom{l}{k}^2$. Thus

$$\begin{aligned} J_l^{(1)} &\leqslant \binom{l}{k}^2 \int_0^1 \cdots \int_0^1 \Bigl| \sum_{[m_1,\dots,m_k], m_{k+1},\dots,m_l} U(m_1) \dots U(m_l) \Bigr|^2 d\alpha_1 \dots d\alpha_k \\ &\leqslant \binom{l}{k}^2 \int_0^1 \cdots \int_0^1 \Bigl| \sum_{[m_1,\dots,m_k]} U(m_1) \dots U(m_k) \Bigr|^2 \times \\ &\qquad \times \Bigl| \sum_{m_{k+1},\dots,m_l} U(m_{k+1}) \dots U(m_l) \Bigr|^2 d\alpha_1 \dots d\alpha_k \\ &\leqslant \binom{l}{k}^2 p_1^{2l-2k-1} \int_0^1 \cdots \int_0^1 \Bigl| \sum_{[m_1,\dots,m_k]} U(m_1) \dots U(m_k) \Bigr|^2 \times \\ &\qquad \times \sum_{m=0}^{p_1-1} |U(m)|^{2l-2k} d\alpha_1 \dots d\alpha_k, \end{aligned} \tag{23}$$

by Hölder's inequality. The integral on the right-hand side is equal to the number of solutions, in integers, of the system of equations

$$\begin{aligned} &(m_1+p_1 a_1)^h + \cdots + (m_k+p_1 a_k)^h - (n_1+p_1 b_1)^h - \cdots - (n_k+p_1 b_k)^h \\ &= (m+p_1 a_{k+1})^h + \cdots + (m+p_1 a_l)^h - (m+p_1 b_{k+1})^h - \cdots - (m+p_1 b_l)^h, \end{aligned} \tag{24}$$

for $h=1,2,\ldots,k$, and for $0\leqslant m, m_j, n_j \leqslant p_1-1$, $0\leqslant j\leqslant k$; $0\leqslant a_j, b_j\leqslant q_1-1$, $0\leqslant j\leqslant l$. We have $m_j\neq m_{j'}$, $n_j\neq n_{j'}$ for $j\neq j'$. This number is also equal to the number of solutions of the system

$$\begin{aligned}&(m_1-m+p_1a_1)^h+\cdots-(n_k-m+p_1b_k)^h\\&=p_1^h(a_{k+1}^h+\cdots+a_l^h-b_{k+1}^h-\cdots-b_l^h).\end{aligned}\tag{25}$$

Let this number be N_l, so that

$$J_l^{(1)}\leqslant\binom{l}{k}^2p_1^{2l-2k-1}N_l.\tag{26}$$

In order to estimate N_l, we define three other numbers. Let $N_k(\lambda_1p_1,\ldots,\lambda_kp_1^k)$ denote the number of solutions of the system of equations

$$(m_1-m+p_1a_1)^h+\cdots-(n_k-m+p_1b_k)^h=\lambda_hp_1^h,\qquad h=1,2,\ldots,k,\tag{27}$$

where the (λ_h) are integers, and the other variables satisfy the same conditions as before. Let T' denote the number of solutions of the system of congruences

$$(m_1-m+p_1a_1)^h+\cdots-(n_k-m+p_1b_k)^h\equiv0(\operatorname{mod}p_1^h),\tag{28}$$

for $h=1,2,\ldots,k$, with $m_j\neq m_{j'}$, $n_j\neq n_{j'}$, for $j\neq j'$. Let $J(q,l,\lambda)$ denote the number of solutions of the system of equations.

$$m_1^h+\cdots+m_l^h=n_1^h+\cdots+n_l^h+\lambda_h,\tag{29}$$

for $h=1,2,\ldots,k$, $0\leqslant m_j, n_j\leqslant q-1$.

Clearly

$$J(q,l,\lambda)=\int_0^1\cdots\int_0^1e^{-2\pi i(\lambda_1\alpha_1+\cdots+\lambda_k\alpha_k)}|S(q)|^{2l}d\alpha_1\ldots d\alpha_k,$$

and

$$J(q,l,\lambda)=|J(q,l,\lambda)|\leqslant J(q,l).\tag{30}$$

Hence

$$\begin{aligned}N_l&\leqslant\sum_{\lambda_1,\ldots,\lambda_k}N_k(\lambda_1p_1,\ldots,\lambda_kp_1^k)\cdot J(q_1,l-k,\lambda)\\&\leqslant J(q_1,l-k)\sum_{|\lambda_h|\leqslant lq_1^h}N_k(\lambda_1p_1,\ldots,\lambda_kp_1^k),\qquad h=1,2,\ldots,k,\\&=J(q_1,l-k)\cdot T'.\end{aligned}\tag{31}$$

To estimate T', we apply Lemma 1 with p_1 in place of p, which is permissible, since $p_1>k$. Now T' is the number of solutions of the system

$$(m_1-m+p_1a_1)^h+\cdots+(m_k-m+p_1a_k)^h$$
$$\equiv(n_1-m+p_1b_1)^h+\cdots+(n_k-m+p_1b_k)^h\ (\mathrm{mod}\, p_1^h),\ h=1,2,\ldots,k,$$

with $m_j\neq m_{j'}$ and $n_j\neq n_{j'}$ for $j\neq j'$, and $0\leqslant m, m_j, n_j\leqslant p_1-1$, $0\leqslant a_j, b_j\leqslant q_1-1$.

This number is also equal to the number of solutions of the system

$$(m_1-m+p_1+p_1a_1)^h+\cdots+(m_k-m+p_1+p_1a_k)^h$$
$$\equiv(n_1-m+p_1+p_1b_1)^h+\cdots+(n_k-m+p_1+p_1b_k)^h\ (\mathrm{mod}\, p_1^h),$$

for $h=1,2,\ldots,k$.

If we put $x_j=m_j-m+p_1+p_1a_j$, $1\leqslant j\leqslant k$, then $x_j\geqslant 0$, and if we choose

$$M=[q_1p_1^{-k+1}]+1$$

in Lemma 1, we have $M>q_1p_1^{-k+1}$, so that $p_1q_1<Mp_1^k$, or $q_1\leqslant Mp_1^{k-1}-1$, and $x_j\leqslant p_1-1-m+p_1q_1\leqslant p_1(q_1+1)-1\leqslant Mp_1^k-1$. Further $x_j\not\equiv x_{j'}\ (\mathrm{mod}\, p_1)$ for $j\neq j'$.

It follows from the Lemma that given (n_j, m, b_j) for $1\leqslant j\leqslant k$, there are at most $k!\,M^kp_1^{\frac{1}{2}k(k-1)}$ solutions $(x_1,\ldots,x_k)$ of the system

$$x_1^h+\cdots+x_k^h\equiv(n_1-m+p_1+p_1b_1)^h+\cdots+(n_k-m+p_1+p_1b_k)^h\ (\mathrm{mod}\, p_1^h),$$

for $h=1,2,\ldots,k$. If we note that m and x_j, $1\leqslant j\leqslant k$, determine (m_j, a_j), $1\leqslant j\leqslant k$, uniquely, and that n_j and m can each assume at most p_1 different values, while b_j can assume at most q_1 different values, we get

$$T'\leqslant p_1(p_1q_1)^k k!\,M^kp_1^{\frac{1}{2}k(k-1)}.$$

Since we have assumed that $p_1\leqslant q^{1/k}$, and $q_1>qp_1^{-1}$, we have $q_1p_1^{-k+1}>1$, and $[q_1p_1^{-k+1}]+1\leqslant 2[q_1p_1^{-k+1}]$. Hence

$$T'\leqslant k!\,2^kq_1^{2k}p_1^{-\frac{1}{2}k(k+1)+2k+1}.$$

If we substitute this in (31), we get

$$N_l\leqslant J(q_1, l-k)\cdot k!\,2^kq_1^{2k}p_1^{2k+1-\frac{1}{2}k(k+1)},$$

which, in conjunction with (26), gives

$$J_l^{(1)}\leqslant p_1^{2l-\frac{1}{2}k(k+1)}q_1^{2k}J(q_1, l-k)\cdot\frac{l^{2k}\cdot 2^k}{k!}$$
$$\leqslant 2l^{2k}q_1^{2k}p_1^{2l-\frac{1}{2}k(k+1)}J(q_1, l-k). \tag{32}$$

Taken together with (22), (18) and (16), this leads to

$$J(q,l) < J(p_1 q_1, l) = J_l^{(1)} + J_l^{(2)} \leqslant 3\, l^{2k} q_1^{2k} p_1^{2l - \frac{1}{2}k(k+1)} J(q_1, l-k),$$

which proves Theorem 1.

By repeated application of inequality (13), we deduce

THEOREM 2 (VINOGRADOV'S MEAN-VALUE THEOREM). *Let r, k, l and q be positive integers, such that*

$$r \geqslant 1, \quad k \geqslant 2, \quad l \geqslant k^2 + kr, \quad q \geqslant (2^k k^{2k})^{\left(1 + \frac{1}{k-1}\right)^{r-1}}. \tag{33}$$

Then

$$J(q,l) < (3\, l^{2k})^r\, 3^{4lr - \frac{1}{2}k(k+1)r}\, q^{2l - \frac{1}{2}k(k+1) + \delta_r}, \tag{34}$$

where

$$\delta_r = \frac{1}{2} k(k+1) \left(1 - \frac{1}{k}\right)^r.$$

PROOF. For $\nu = 1, 2, \ldots, r$, we define the integer q_ν and the prime p_ν, by the following conditions:

$$q_0 = q, \quad q_\nu = [q_{\nu-1} p_\nu^{-1}] + 1, \tag{35}$$

and

$$\tfrac{1}{2} q_{\nu-1}^{1/k} \leqslant p_\nu \leqslant q_{\nu-1}^{1/k}, \tag{36}$$

which accord with the conditions (14) and (15). Then we have the inequalities

$$q_{\nu-1} < p_\nu q_\nu, \quad q^{\left(1 - \frac{1}{k}\right)^\nu} < q_\nu < 3^\nu q^{\left(1 - \frac{1}{k}\right)^\nu}, \quad p_\nu > k^2. \tag{37}$$

The first of these is immediate from (35). To prove the second, we observe that

$$q_\nu > \frac{q_{\nu-1}}{p_\nu} \geqslant q_{\nu-1}^{1 - \frac{1}{k}} \geqslant q^{\left(1 - \frac{1}{k}\right)^\nu},$$

and

$$q_\nu \leqslant \frac{q_{\nu-1}}{p_\nu} + 1 \leqslant 2 q_{\nu-1}^{1 - \frac{1}{k}} + 1 < 3 q_{\nu-1}^{1 - \frac{1}{k}} < 3^\nu q^{\left(1 - \frac{1}{k}\right)^\nu}.$$

To prove the third, we observe that

$$p_\nu \geqslant \tfrac{1}{2} q_{\nu-1}^{1/k} > \tfrac{1}{2} q^{\frac{1}{k}\left(1 - \frac{1}{k}\right)^{\nu-1}} \geqslant k^2,$$

because of our assumption on q in (33).

Now, in view of the definition of p_ν and q_ν, and of the assumptions

$$l \geqslant k^2 + kr, \qquad q_{r-1} \geqslant 2^{-k}(2k)^{2k},$$

the latter being a consequence of the assumption made on q (we note that q_ν is a decreasing sequence), and of (37), we can apply the reduction formula given by (13) r times, so as to get the inequality

$$J(q,l) < (3\,l^{2k})^r\,(q_1\cdots q_r)^{2k}\,(p_1\cdots p_r)^{2l-\frac{1}{2}k(k+1)}\cdot(p_2\cdot p_3^2\cdots p_r^{r-1})^{-2k}\,J(q_r, l-kr).$$

From (37), however, we have

$$q_1\cdots q_r < p_2\cdot p_3^2\cdots p_r^{r-1}\,q_r^r,$$

and since $q_\nu = [q_{\nu-1}p_\nu^{-1}]+1$, and $(q_{\nu-1}/p_\nu) > 1$, we have $q_\nu < (2q_{\nu-1}/p_\nu)$, so that

$$p_1\cdots p_r < 2\frac{q}{q_1}\cdot 2\frac{q_1}{q_2}\cdots 2\frac{q_{r-1}}{q_r} = 2^r\frac{q}{q_r} < 2^r q\cdot q^{-\left(1-\frac{1}{k}\right)^r} < 3^r q^{1-\left(1-\frac{1}{k}\right)^r},$$

while trivially we have $J(q_r, l-kr) \leqslant q_r^{2(l-kr)}$. Hence

$$J(q,l) < (3\,l^{2k})^r\cdot 3^{\left(2l-\frac{1}{2}k(k+1)\right)r}\cdot q_r^{2l}\cdot q^{\left(2l-\frac{1}{2}k(k+1)\right)\left(1-\left(1-\frac{1}{k}\right)^r\right)}, \tag{38}$$

and since

$$q_r^{2l} < 3^{2lr} q^{\left(1-\frac{1}{k}\right)^r\cdot 2l},$$

we have

$$J(q,l) < (3\,l^{2k})^r\,3^{4lr-\frac{1}{2}k(k+1)r}\,q^{2l-\frac{1}{2}k(k+1)+\delta_r}, \tag{38'}$$

as claimed in the theorem.

The estimate given in Theorem 2 is valid for

$$q \geqslant (2^k k^{2k})^{\left(1+\frac{1}{k-1}\right)^{r-1}},$$

where r is an integer $\geqslant 1$. We shall now put it in a form which is valid without any special restriction on q, by making an appropriate choice of r. We do not attempt the best possible form.

THEOREM 3. *Let k, l, and q be positive integers, such that*

$$k \geqslant 2, \qquad l \geqslant k^2 + 4k^2\log k, \qquad q \geqslant 2.$$

Then there exists an absolute numerical constant c, such that

$$J(q,l) \leqslant e^{clk\log^2 k}\,q^{2l-\frac{1}{2}k(k+1)+\frac{1}{2}}. \tag{39}$$

PROOF. We note that (38) and (38′) hold under the assumption $q_{r-1} \geqslant 2^{-k}(2k)^{2k}$. If q is such that $q < 2^{-k}(2k)^{2k}$, then trivially we have

$$J(q, l) \leqslant q^{2l} < e^{4kl \log 2k},$$

and since $2l - \frac{1}{2}k(k+1) + \frac{1}{2} > 0$, by hypothesis, it follows that (39) is valid.

If $q \geqslant 2^{-k}(2k)^{2k}$, then there are two possibilities. In the notation of Theorem 2, either (i) there exists an integer r, in the range $1 \leqslant r \leqslant 4k \log k$, such that

$$q_{r-1} \geqslant 2^{-k}(2k)^{2k}, \qquad q_r < 2^{-k}(2k)^{2k},$$

or (ii) we have

$$q_r \geqslant 2^{-k}(2k)^{2k},$$

for every r in the range $1 \leqslant r \leqslant 4k \log k$.

In the first case, we can apply Theorem 1 r times, provided that $l \geqslant k^2 + kr$, which we have assumed. Hence, as in (38), we have

$$J(q, l) < (3\,l^{2k})^r\, 3^{\left(2l - \frac{1}{2}k(k+1)\right)r} (2^k k^{2k})^{2l}\, q^{\left(2l - \frac{1}{2}k(k+1)\right)\left(1 - \left(1 - \frac{1}{k}\right)^r\right)},$$

and, because of the condition on r, (39) is again valid, if we note that $k \log l \leqslant l \log k$, for $l \geqslant k^2 \geqslant 4$.

In the second case, we choose

$$r = [k \log(k^2 + k)] + 1,$$

so that $\dfrac{r}{k} \geqslant \log(k^2 + k)$, which implies that

$$r \log \frac{k}{k-1} \geqslant \log(k^2 + k),$$

since $k \geqslant 2$. Hence $\delta_r = \dfrac{1}{2}k(k+1)\left(1 - \dfrac{1}{k}\right)^r \leqslant \dfrac{1}{2}$. Since $q_r \geqslant 2^{-k}(2k)^{2k}$, we can apply, as before, Theorem 1 r times, noting that $r \leqslant 4k \log k$, and we see, as in (38)′, that (39) is again valid.

§ 4. Vinogradov's inequality. We shall use Theorem 3 to estimate trigonometric sums of the form

$$C = \sum_{n=P+1}^{P+Q} e^{2\pi i F(n)},$$

where P and Q are integers, and F a real-valued function with a certain number of continuous derivatives in the interval $P + 1 \leqslant x \leqslant P + Q$.

We require a preliminary

LEMMA 2 (VINOGRADOV). *Let $\varphi(n)$ be a real-valued function of n (where n is an integer) defined for $M \leqslant n \leqslant M+N-1$, M and N being integers, with $N>1$, such that*

$$\delta \leqslant \varphi(n+1)-\varphi(n) \leqslant a\delta, \qquad M \leqslant n \leqslant M+N-2, \tag{40}$$

where $\delta>0$. Let $W>0$. Then the number of values of n for which $\|\varphi(n)\| \leqslant W\delta$ is less than

$$(Na\delta+1)(2W+1), \tag{41}$$

where $\|x\|$ denotes the distance of x from the nearest integer.

PROOF. The estimate is trivial if $W\delta > \frac{1}{2}$.

For a given real number ξ, and a given integer h, there cannot be more than one value of n, in the interval considered, for which

$$\xi+h<\varphi(n) \leqslant \xi+h+\delta,$$

because of the hypothesis (40). Thus, if G denotes the number of values of n satisfying this condition for a given ξ, and for *some* integer h, then $G \leqslant h_2-h_1+1$, where h_1 and h_2 are the least, and the greatest, values of h in question. Since

$$\varphi(M) \leqslant \xi+h_1+\delta, \quad \text{and} \quad \xi+h_2<\varphi(M+N-1),$$

we have

$$h_2-h_1-\delta<\varphi(M+N-1)-\varphi(M) \leqslant (N-1)a\delta,$$

and therefore

$$G<(N-1)a\delta+\delta+1 \leqslant Na\delta+1,$$

since $a \geqslant 1$ by definition. The result now follows upon dividing the interval $[-W\delta, W\delta]$ into $[2W+1]$ intervals each of length less than δ, and applying the above inequality.

THEOREM 4 (VINOGRADOV'S INEQUALITY). *Let P and Q be positive integers, $Q \geqslant 2$, and let F be a real-valued function with a continuous derivative of order $k+1$ in the interval $P+1 \leqslant x \leqslant P+Q$, where $k \geqslant 7$. Let λ be a real number, such that*

$$\lambda \leqslant \left|\frac{F^{(k+1)}(x)}{(k+1)!}\right| \leqslant 2\lambda, \tag{42}$$

for $P+1 \leqslant x \leqslant P+Q$, and let

$$\frac{1}{\lambda^{\frac{1}{4}}} \leqslant Q \leqslant \frac{1}{\lambda}, \tag{43}$$

(so that $0<\lambda\leqslant\frac{1}{2}$). Then

$$\left|\sum_{n=P+1}^{P+Q} e^{2\pi i F(n)}\right| < K\cdot e^{c_1 k\log^2 k} Q^{1-\rho}, \tag{44}$$

where K and c_1 are absolute constants ($c_1=\frac{1}{2}c$, where c is defined in Theorem 3), and

$$\rho=\frac{1}{70k^2\log k}. \tag{45}$$

PROOF. The first step in the proof is to show that if

$$C=\sum_{n=P+1}^{P+Q} e^{2\pi i F(n)},$$

then we have the inequality

$$|C|\leqslant q^{-1}\sum_{n=P+1}^{P+Q-q}|T(n)|+q, \tag{46}$$

where q is an integer such that $1\leqslant q<Q$, and

$$T(n)=\sum_{m=1}^{q} e^{2\pi i\{F(m+n)-F(n)\}},$$

where $P+1\leqslant n\leqslant P+Q-q$. Actually we shall choose

$$q=\left[\lambda^{\frac{\eta-1}{k+1}}\right],\qquad \eta=\frac{1}{6k^2\log k}.$$

Since $0<\lambda\leqslant\frac{1}{2}$, $k\geqslant 7$, we have $q\geqslant 1$, and since $\lambda^{-\frac{1}{4}}\leqslant Q$, $k\geqslant 7$, we have $q\leqslant[Q^{4/(k+1)}]<Q$.

To prove (46), we consider

$$\begin{aligned} q|C| &= \left|\sum_{m=1}^{q}\sum_{n=P+1}^{P+Q} e^{2\pi i F(n)}\right| \\ &\leqslant \left|\sum_{m=1}^{q}\sum_{n=P+1+m}^{P+Q-q+m} e^{2\pi i F(n)}\right| + \sum_{m=1}^{q} q \\ &= \left|\sum_{m=1}^{q}\sum_{n=P+1}^{P+Q-q} e^{2\pi i F(m+n)}\right| + q^2 \\ &= \left|\sum_{n=P+1}^{P+Q-q}\sum_{m=1}^{q} e^{2\pi i F(m+n)}\right| + q^2 \\ &\leqslant \sum_{n=P+1}^{P+Q-q}|T(n)|+q^2, \end{aligned}$$

and this gives (46). By Hölder's inequality, we get

$$q|C| \leqslant Q^{1-\frac{1}{2l}} \left\{ \sum_{n=P+1}^{P+Q-q} |T(n)|^{2l} \right\}^{1/(2l)} + q^2, \tag{47}$$

where l is any positive integer.

The second step in the proof consists in showing that the sum involving $T(n)$ on the right-hand side of (47) can be majorized by an integral of the form

$$\int_0^1 \cdots \int_0^1 |S(q)|^{2l} \, d\alpha_1 \ldots d\alpha_k,$$

where $S(q)$ is defined as in (1).

Since $P+1 \leqslant n \leqslant P+Q-q$ by choice, we have, for $1 \leqslant m \leqslant q$,

$$F(m+n) - F(n) = A_1 m + \cdots + A_k m^k + 2\lambda\theta q^{k+1},$$

where $A_r = A_r(n) = \dfrac{F^{(r)}(n)}{r!}$, and $|\theta| \leqslant 1$. Now define the k-dimensional region Ω_n, by the property that $\alpha = (\alpha_1, \ldots, \alpha_k) \in \Omega_n$, if and only if

$$|\alpha_r - A_r| \leqslant \frac{1}{2} \cdot \frac{1}{q^r} (\lambda q^{k+1}), \qquad r = 1, \ldots, k.$$

We call the point $(A_1, \ldots, A_k)$ the *centre* of Ω_n.

It is clear that for $1 \leqslant m \leqslant q$, and $\alpha \in \Omega_n$, we have

$$\begin{aligned}
&|e^{2\pi i \{F(m+n) - F(n)\}} - e^{2\pi i(\alpha_k m^k + \cdots + \alpha_1 m)}| \\
&\leqslant 2\pi |F(m+n) - F(n) - \alpha_k m^k - \cdots - \alpha_1 m| \\
&\leqslant 2\pi \sum_{r=1}^{k} |\alpha_r - A_r| m^r + 4\pi\lambda q^{k+1} \\
&\leqslant \pi\lambda q^{k+1} \sum_{r=1}^{k} \left(\frac{m}{q}\right)^r + 4\pi\lambda q^{k+1} \\
&\leqslant \pi\lambda q^{k+1}(k+4) \leqslant 2\pi k\lambda q^{k+1}, \quad \text{since} \quad k \geqslant 7.
\end{aligned}$$

Hence

$$|T(n)| \leqslant |S(q)| + 2\pi k\lambda q^{k+2},$$

which implies that

$$|T(n)|^{2l} \leqslant 2^{2l} |S(q)|^{2l} + (4\pi k\lambda q^{k+2})^{2l}.$$

Integrating over the region Ω_n, and dividing by the volume (= measure) of Ω_n, we get

$$|T(n)|^{2l} \leqslant 2^{2l} q^{\frac{1}{2}k(k+1)} (\lambda q^{k+1})^{-k} \int \cdots \int_{\Omega_n} |S(q)|^{2l} \, d\alpha_1 \ldots d\alpha_k + (4\pi k\lambda q^{k+2})^{2l}. \tag{48}$$

Now $S(q)$ is a periodic function in each of the variables $\alpha_1, \ldots, \alpha_k$, with period 1. If we denote by (Ω_n+g) the set of points $(\alpha_1+g_1, \ldots, \alpha_k+g_k)$, where $(\alpha_1, \ldots, \alpha_k) \in \Omega_n$, and $g_1, \ldots, g_k$ are integers (positive, negative, or zero), then the integral of $|S(q)|^{2l}$ over Ω_n equals its integral over (Ω_n+g).

Consider the set

$$\Omega'_n = \bigcup_g (\Omega_n+g),$$

the union being taken over all $g=(g_1, \ldots, g_k)$, and denote by D_n the intersection of Ω'_n with the k-dimensional unit cube defined by $0 \leqslant x_j < 1$, $j=1,2,\ldots,k$. Let $\chi(D_n)=\chi(D_n,\alpha)$ denote the characteristic function of D_n, which equals 1 if $\alpha \in D_n$ and 0 otherwise. Then, because of our choice of q, we have $\lambda q^k < 1$, which ensures that the (Ω_n+g) do not overlap, and (48) gives

$$|T(n)|^{2l} \leqslant 2^{2l} q^{\frac{1}{2}k(k+1)} (\lambda q^{k+1})^{-k} \int_0^1 \cdots \int_0^1 |S(q)|^{2l} \chi(D_n) d\alpha_1 \ldots d\alpha_k + \\ + (4\pi k \lambda q^{k+2})^{2l},$$

which implies that

$$\sum_{n=P+1}^{P+Q-q} |T(n)|^{2l} \leqslant 2^{2l} q^{\frac{1}{2}k(k+1)} (\lambda q^{k+1})^{-k} \int_0^1 \cdots \int_0^1 |S(q)|^{2l} \sum_{n=P+1}^{P+Q-q} \chi(D_n) d\alpha_1 \ldots d\alpha_k \\ + (Q-q)(4\pi k \lambda q^{k+2})^{2l}. \tag{49}$$

Now $\sum_{n=P+1}^{P+Q-q} \chi(D_n,\alpha)$ equals the number of regions D_n, with $P+1 \leqslant n \leqslant P+Q-q$, to which α belongs. If α belongs to a given region D_n, and also to $D_{n'}$, and $P+1 \leqslant n, n' \leqslant P+Q-q$, then the kth coordinates of the "centres" of the regions Ω_n and $\Omega_{n'}$, namely $A_k(n)$, and $A_k(n')$, satisfy the condition

$$\|A_k(n) - A_k(n')\| \leqslant q^{-k}(\lambda q^{k+1}) = \lambda q, \tag{50}$$

where $\|x\|$ denotes the distance of x from the nearest integer. Hence the number of regions D_n to which α belongs does not exceed the number of numbers n in the closed interval $[P+1, P+Q-q]$ for which (50) is valid. We shall prove that this number is $\leqslant 4kq$, a bound which is independent of n', by appealing to Lemma 2, with

$$\varphi(n) = A_k(n) - A_k(n'), \quad M=P+1, \quad N=Q-q>1, \quad a=2, \quad \delta=\lambda(k+1),$$

$$W = \frac{q}{k+1}.$$

(If $Q-q=1$, the assertion is trivial.) We observe that

$$\varphi(n+1)-\varphi(n) = \frac{1}{k!}\{F^{(k)}(n+1)-F^{(k)}(n)\} = \frac{F^{(k+1)}(\xi)}{k!},$$

where ξ is a number such that $n<\xi<n+1$. Hypothesis (42) ensures that $\frac{F^{(k+1)}(x)}{(k+1)!}$ is of the same sign throughout the interval $P+1\leqslant x \leqslant P+Q$. We can suppose, by taking conjugates in (44) if necessary, that $F^{(k+1)}(x)$ is positive throughout. It follows from the Lemma that the number of values of n in the interval $P+1\leqslant n\leqslant P+Q-q$, for which

$$\|A_k(n)-A_k(n')\|\leqslant \lambda q,$$

does not exceed

$$(2\lambda(k+1)Q+1)\left(\frac{2q}{k+1}+1\right) \leqslant (2k+3)\left(\frac{2q}{k+1}+1\right) \leqslant 3k\left(\frac{q}{4}+q\right) < 4kq,$$

since $Q\leqslant \frac{1}{\lambda}$, $q\geqslant 1$, $k\geqslant 7$.

Reverting to (49), and using the definition of $J(q,l)$, we obtain

$$\sum_{n=P+1}^{P+Q-q} |T(n)|^{2l}\leqslant 2^{2l}\cdot q^{\frac{1}{2}k(k+1)}\cdot(\lambda q^{k+1})^{-k}\cdot 4kq\,J(q,l)+(Q-q)(4\pi k\lambda q^{k+2})^{2l}.$$

In conjunction with (47), this gives

$$|C|\leqslant q^{-1}Q^{1-\frac{1}{2l}}\left\{\sum_{n=P+1}^{P+Q-q}|T(n)|^{2l}\right\}^{1/(2l)}+q$$

$$\leqslant q^{-1}Q^{1-\frac{1}{2l}}\left\{2^{2l}\lambda^{-k}q^{-\frac{1}{2}k(k+1)}4kq\,J(q,l)+(Q-q)(4\pi k\lambda q^{k+2})^{2l}\right\}^{1/(2l)}+q.$$

We now choose l as in Theorem 3, namely

$$l=[k^2+4k^2\log k]+1,$$

and use the estimate for $J(q,l)$ obtained in that theorem, so as to obtain the inequality

$$|C|\leqslant 2q^{-1}Q^{1-\frac{1}{2l}}\left\{2^{2l}\cdot\lambda^{-k}\cdot q^{-\frac{1}{2}k(k+1)}\cdot 4kq\cdot J(q,l)\right\}^{1/(2l)}+$$

$$+2q^{-1}\cdot Q^{1-\frac{1}{2l}}\cdot Q^{1/(2l)}(4\pi k\lambda q^{k+2})+q$$

$$\leqslant 4Q^{1-\frac{1}{2l}}\left\{4k\cdot\lambda^{-k}\cdot q^{\frac{3}{2}-k(k+1)}\cdot e^{clk\log^2 k}\right\}^{1/(2l)}+8\pi k\lambda q^{k+1}Q+q.$$

By the definition of q, we have

$$q \leqslant Q^{4/(k+1)}, \qquad \lambda q^{k+1} \leqslant \lambda^{\eta} \leqslant Q^{-\eta},$$

while $q \geqslant \frac{1}{2}\lambda^{\frac{\eta-1}{k+1}}$, since $\lambda^{\frac{\eta-1}{k+1}} \geqslant 1$, so that

$$\lambda q^{k+1} \geqslant 2^{-(k+1)}\lambda^{\eta} \geqslant 2^{-(k+1)} Q^{-4\eta}.$$

Hence

$$|C| \leqslant c_2 e^{c_1 k \log^2 k} \cdot Q^{1-\frac{1}{2l}} \cdot Q^{\frac{2\eta k}{l}} \cdot Q^{\frac{3}{(k+1)l}} + 8\pi k Q^{1-\eta} + Q^{\frac{4}{(k+1)}},$$

where c_1, c_2 are absolute constants. For $k \geqslant 7$, we have

$$\tfrac{1}{2} - 3(k+1)^{-1} - 2\eta k \geqslant \tfrac{1}{2} - \tfrac{3}{8} - \tfrac{1}{21} \geqslant \tfrac{1}{14},$$

and $l < 5k^2 \log k$, and inequality (44) claimed in Theorem 4 follows.

We shall now cast Theorem 4 in a form more convenient for applications.

THEOREM 5. *If $F(x)$ is a real-valued function with a continuous derivative of order $(k+1)$, where $k \geqslant 7$, in $P+1 \leqslant x \leqslant P+N$, where $N \leqslant Q$, Q real, and*

$$\lambda \leqslant \left|\frac{F^{(k+1)}(x)}{(k+1)!}\right| \leqslant 2\lambda, \tag{51}$$

$$\frac{1}{\lambda^{\frac{1}{3}}} \leqslant Q \leqslant \frac{1}{\lambda},$$

then

$$\left|\sum_{n=P+1}^{P+N} e^{2\pi i F(n)}\right| < c_3 e^{c_1 k \log^2 k} Q^{1-\rho}, \tag{52}$$

where $\rho = \dfrac{1}{70k^2 \log k}$, *and c_1, c_3 are absolute constants.*

PROOF. If $\dfrac{1}{\lambda^{\frac{1}{4}}} \leqslant N$, then the proof is immediate, for we have only to apply Theorem 4 with N in place of the Q there. If $\lambda^{-\frac{1}{4}} > N$, then we have

$$\left|\sum_{n=P+1}^{P+N} e^{2\pi i F(n)}\right| \leqslant N < \frac{1}{\lambda^{\frac{1}{4}}} \leqslant Q^{\frac{3}{4}} \leqslant Q^{1-\rho},$$

and the required result follows.

§ 5. Estimation of sections of $\zeta(s)$ in the critical strip. We can apply Theorems 4 and 5 to estimate the sum

$$\sum_{a<n\leqslant b} \frac{1}{n^{\sigma+it}}, \quad \text{for} \quad 0<\sigma<1, \quad t\geqslant 1, \quad a<b\leqslant 2a.$$

For this purpose we choose

$$F(x) = -\frac{t\log x}{2\pi}, \tag{53}$$

so that we have

$$F^{(k+1)}(x) = \frac{(-1)^{k+1}k!\,t}{2\pi x^{k+1}},$$

and $(-1)^{k+1}F^{(k+1)}(x)$ is monotone decreasing in the interval $0<a\leqslant x \leqslant b\leqslant 2a$. We can divide the interval (a,b) into not more than $(k+1)$ sub-intervals, in each of which condition (51) holds with a λ depending on the particular sub-interval. Since

$$\frac{t}{2\pi(k+1)(2a)^{k+1}} \leqslant \left|\frac{F^{(k+1)}(x)}{(k+1)!}\right| \leqslant \frac{t}{2\pi(k+1)a^{k+1}},$$

throughout the interval $a<x\leqslant 2a$, it follows that every such λ may be chosen to satisfy the inequalities

$$\frac{t}{2\pi(k+1)(2a)^{k+1}} \leqslant \lambda \leqslant \frac{t}{4\pi(k+1)a^{k+1}}. \tag{54}$$

We choose $Q=a$, $\log a>2\log^{\frac{1}{2}} t$, in Theorem 5, and

$$k = \left[\frac{\log t}{\log a}\right] + 1,$$

so that

$$Q < \frac{a^{k+1}}{t} \leqslant Q^2. \tag{55}$$

Because of (54) and (55), we have $\lambda < \frac{1}{Q}$, while

$$\lambda \geqslant \frac{1}{Q^3}, \tag{56}$$

if $Q\geqslant 2^{k+2}\pi(k+1)$; that is, if

$$\log a \geqslant \left(\frac{\log t}{\log a} + 3\right)\log 2 + \log\pi + \log\left(\frac{\log t}{\log a} + 2\right), \tag{57}$$

because of our choice of k. Since $\log a > 2\log^{\frac{1}{2}} t$, (57) holds if t is large enough, and therefore (56) as well.

If we *assume* further that $a \leqslant t^{\frac{1}{6}}$, then $\log a \leqslant \frac{1}{6}\log t$, so that $k \geqslant 7$.

By applying Theorem 5 to each of the sub-intervals of $(a, b]$, at most $k+1$ in number, we obtain, after a partial summation, the estimate

$$\sum_{a<n\leqslant b} \frac{1}{n^{\sigma+it}} = O(k \cdot e^{c_1 k \log^2 k} a^{1-\rho-\sigma}), \tag{58}$$

where $0<\sigma<1$, $\rho=(70k^2\log k)^{-1}$ (see Notes on Ch. III, §§ 2—3).

Now suppose a is such that

$$k \log k < \varepsilon \log^{\frac{1}{3}} a, \tag{59}$$

for a sufficiently small $\varepsilon>0$. Because of our choice of k, it follows that this condition is equivalent to

$$\log a > A(\log t \log\log t)^{\frac{3}{4}}, \tag{60}$$

with $A = A(\varepsilon)$, for t sufficiently large. For if $\log a \leqslant c_3(\log t \log\log t)^{\frac{3}{4}}$, for some constant c_3, then

$$k > \frac{\log t}{\log a} \geqslant \frac{\log^{\frac{1}{4}} t}{c_3(\log\log t)^{\frac{3}{4}}},$$

which implies that $\log k \geqslant \frac{1}{4}\log\log t - \log c_3 - \frac{3}{4}\log\log\log t > \frac{1}{5}\log\log t$, if $t \geqslant t_0(c_3)$. Hence, by (59),

$$\varepsilon \log^{\frac{1}{3}} a > k \log k > \frac{\log t}{\log a} \cdot \log k > \frac{\log t}{\log a} \cdot \frac{1}{5} \log\log t,$$

that is

$$\log a > \left(\frac{1}{5\varepsilon}\right)^{\frac{3}{4}} (\log t \log\log t)^{\frac{3}{4}},$$

which implies that $c_3 > \left(\frac{1}{5\varepsilon}\right)^{\frac{3}{4}}$. Hence if we take a number less than $\left(\frac{1}{5\varepsilon}\right)^{\frac{3}{4}}$, say $\left(\frac{1}{6\varepsilon}\right)^{\frac{3}{4}}$, then

$$\log a > \left(\frac{1}{6\varepsilon}\right)^{\frac{3}{4}} (\log t \log\log t)^{\frac{3}{4}},$$

which proves (60). Obviously this implies the previous assumption that $\log a > 2\log^{\frac{1}{2}} t$, if t is sufficiently large.

Conversely if (60) holds, for a sufficiently large A, then

$$k < \frac{2\log t}{\log a} < \frac{2}{A} \cdot \frac{\log^{\frac{1}{4}} t}{(\log\log t)^{\frac{3}{4}}},$$

and $\log k < \frac{1}{2}\log\log t$, because $k < \dfrac{2\log t}{\log a} < \log^{\frac{1}{2}} t$. Hence

$$k\log k < \frac{1}{A}\cdot\log^{\frac{1}{4}} t\cdot\log\log^{\frac{1}{4}} t < \frac{1}{A}\left(\frac{1}{A}\log a\right)^{\frac{1}{3}},$$

which proves (59).

If we assume (60), which implies (59), then from (58) we get

$$\sum_{a<n\leqslant b}\frac{1}{n^{\sigma+it}} = O\left(\frac{\log t}{\log a}\cdot a^{1-\sigma}\cdot\exp\left(c_1 k\log^2 k - \frac{\log a}{70k^2\log k}\right)\right)$$

$$= O\left(\frac{\log t}{\log a}\cdot a^{1-\sigma}\cdot\exp\left(\frac{(70c_1\varepsilon^3-1)\log a}{70k^2\log k}\right)\right),$$

because of (59). Noting once again that $k<2\dfrac{\log t}{\log a}$ on the one hand, and $\log k<\frac{1}{2}\log\log t$ on the other, we obtain

$$\sum_{a<n\leqslant b}\frac{1}{n^{\sigma+it}} = O(\log t\cdot e^{(1-\sigma)\log a}\cdot e^{-c_2\log^3 a/(\log^2 t\log\log t)}), \tag{61}$$

where $c_2>0$, if $\varepsilon>0$, and sufficiently small, t being sufficiently large. Because of (60), and because $\log a\leqslant\frac{1}{6}\log t$, this implies that

$$\sum_{a<n\leqslant b}\frac{1}{n^{\sigma+it}} = O\left(\log t\cdot e^{(1-\sigma)\log a}\cdot\exp\left(\frac{-B\log a\cdot\log\log^{\frac{1}{3}} t}{\log^{\frac{1}{3}} t}\right)\right), \tag{62}$$

where $B=B(A)>0$. This is the type of estimate which we set out to prove, and which we can use to obtain more precise information on the zero-free region of $\zeta(s)$ than was possible in Chapter III.

§ 6. Chudakov's theorem. In the proof of the Weyl-Hardy-Littlewood theorem (Ch. III, Theorem 3), we have seen (Ch. III, (55)) that

$$\zeta(s) = \sum_{1\leqslant n\leqslant t^{2/r}}\frac{1}{n^{\sigma+it}} + O(1),$$

uniformly for $1-\dfrac{1}{R}\leqslant\sigma<1$, where $R=2^{r-1}$, and r is an integer, such that $6\leqslant r\leqslant\log\log t$. If we split up this sum into

$$\left(\sum_{1\leqslant n\leqslant\alpha} + \sum_{\alpha<n\leqslant t^{2/r}}\right)\frac{1}{n^{\sigma+it}},$$

where α is such that

$$\log\alpha = A_1(\log t\log\log t)^{\frac{3}{4}}, \qquad A_1>0, \tag{63}$$

with *a suitably chosen* A_1, and if

$$1-\sigma < \frac{A_2(\log\log t)^{\frac{1}{2}}}{\log^{\frac{1}{2}} t} = 1-\sigma_0, \qquad A_2>0, \tag{64}$$

then

$$\sum_{1\leqslant n\leqslant\alpha} \frac{1}{n^{\sigma+it}} = O\left(\sum_{n\leqslant\alpha} \frac{1}{n^{\sigma_0}}\right) = O\left(\frac{\alpha^{1-\sigma_0}}{1-\sigma_0}\right) = O\left(e^{(1-\sigma_0)\log\alpha}\cdot\frac{\log^{\frac{1}{2}} t}{(\log\log t)^{\frac{1}{2}}}\right)$$

$$= O\left(\exp(A_3(\log\log t)^{\frac{5}{4}}\log^{\frac{1}{4}} t)\cdot\frac{\log^{\frac{1}{2}} t}{(\log\log t)^{\frac{1}{2}}}\right), \tag{65}$$

with $A_3 = A_1 \cdot A_2$.

If R is so chosen that $\dfrac{1}{R} \geqslant 1-\sigma_0$, such a choice being consistent with the previous condition $6 \leqslant r \leqslant \log\log t$, then because of our choice of α in (63), we can apply the estimate (62) to each of the $O(\log t)$ sums of the type

$$\sum_{\alpha<n\leqslant 2\alpha}, \qquad \sum_{2\alpha<n\leqslant 4\alpha}, \ldots,$$

into which $\sum_{\alpha<n\leqslant t^{2/r}}$ is split up, provided that $r \geqslant 12$.

We choose A_1 in (63), and A_2 in (64), such that $A_2 < B$, where B is given by (62), while $A_1 > A$, where A is given by (60). We then obtain

$$\sum_{\alpha<n\leqslant t^{2/r}} n^{-\sigma-it} = O\left(\log^2 t\cdot\exp\left(\frac{-c\log\alpha\cdot\log\log^{\frac{1}{2}} t}{\log^{\frac{1}{2}} t}\right)\right),$$

where $c = B - A_2 > 0$, because of our choice of $1-\sigma$ in (64). And because of our choice of α in (63), we deduce that

$$\sum_{\alpha<n\leqslant t^{2/r}} n^{-\sigma-it} = O(\log^2 t\cdot e^{-c'\log^{1/4} t\cdot\log\log^{5/4} t}) = O(1), \tag{66}$$

since $c' > 0$. By (66), (65), and (64), in conjunction with (55) of Chapter III, we have

$$\zeta(s) = O(e^{A_4(\log\log t)^{5/4}(\log t)^{1/4}}) \tag{67}$$

for

$$1-\sigma < A_2\frac{(\log\log t)^{\frac{1}{2}}}{\log^{\frac{1}{2}} t}, \qquad t > t_0.$$

If we appeal to Lemma 3 of Chapter III, with

$$\theta(t) = \frac{A_2(\log\log t)^{\frac{1}{2}}}{\log^{\frac{1}{2}} t},$$

$$\varphi(t) = A_4\log^{\frac{1}{4}} t\cdot(\log\log t)^{\frac{5}{4}},$$

we obtain the following

THEOREM 6 (CHUDAKOV). *There exists a constant* $A_1 > 0$, *such that* $\zeta(s)$ *has no zeros in the region*

$$\sigma \geqslant 1 - \frac{A_1}{\log^{\frac{3}{4}} t (\log\log t)^{\frac{3}{4}}}, \qquad t \geqslant 3. \tag{68}$$

§ 7. Approximation of $\pi(x)$. In view of Theorem 6, we can choose

$$\eta(t) = \frac{A_1}{(\log t)^{\frac{3}{4}} (\log\log t)^{\frac{3}{4}}}, \qquad t \geqslant 3, \qquad A_1 > 0,$$

so that the conditions imposed on η in Theorem 5 of Chapter III are satisfied (in particular, A_1 has to be suitably chosen).

Let $t = t_0 = t_0(x)$ be the unique solution of the equation $\eta(t)\log x = \log t$, the solution being unique since η is a decreasing function. Then

$$\omega(x) = \min_{t \geqslant 1} (\eta(t)\log x + \log t) > \log t_0 \qquad (x > 1).$$

This is obvious for $t > t_0$; and holds for $t \leqslant t_0$, since $\eta(t) \geqslant \eta(t_0)$. But

$$\log t_0 = \frac{A_1 \log x}{(\log t_0)^{\frac{3}{4}} (\log\log t_0)^{\frac{3}{4}}},$$

so that $\log t_0 = (A_1 \log x)^{\frac{4}{7}} (\log\log t_0)^{-\frac{3}{7}}$, which implies that

$$\log\log t_0 < \log\log t_0 + \tfrac{3}{7} \log\log\log t_0 = \tfrac{4}{7} \log\log x + \tfrac{4}{7} \log A_1 < c_1 \log\log x,$$

for $x > x_0$. Hence $\omega(x) > c_2 (\log x)^{\frac{4}{7}} (\log\log x)^{-\frac{3}{7}}$ for $x > x_0$. If we now appeal to Theorem 7 of Chapter III, we obtain

THEOREM 7. *As* $x \to \infty$, *we have*

$$\pi(x) - \operatorname{li} x = O\left(x e^{-c(\log x)^{\frac{4}{7}} (\log\log x)^{-\frac{3}{7}}}\right),$$

for a positive, absolute constant c.

Notes on Chapter IV

As a general reference see I. M. Vinogradov, *The method of trigonometrical sums in the theory of numbers*, Trav. Inst. Math. Steklov, 10 (1937); second edition, 23 (1947); English translation, Interscience Publishers, London-New York, 1955.

There is an exposition in Titchmarsh's *Zeta-function*, Ch. VI. On page 110, however, there is an error in the simultaneous assumption that $\lambda > 1$, and $\lambda^{-\frac{1}{4}} \leqslant \lambda^{-1}$.

Other general references are L. K. Hua, *Additive Primzahltheorie*. Leipzig (1959); *Abschätzungen von Exponentialsummen und ihre Anwendung in der Zahlentheorie*, Enzyklopädie der mathematischen Wissenschaften (Neuausgabe), Erster Band, Zweiter Teil, Heft 13, Teil 1, 1 – 123; Leipzig (1959); A. Walfisz, *Weylsche Exponentialsummen in der neueren Zahlentheorie*, Berlin 1963.

§§ 1—2. Vinogradov's method was developed by him in his papers *Mat. Sbornik*, 42 (1935), 521—30; ibid. (1) 43 (1936), 9—19; ibid. (1) 43 (1936), 175—188. In our exposition we have used the simplifications effected in later years.

For the properties of symmetric polynomials used in the solution of a system of equations such as (6), see, for instance, B. L. van der Waerden, *Moderne Algebra*, § 26.

§ 3. The proofs of Lemma 1, and of Theorem 1, are due to Karacuba and Korobov, *Doklady Akad. Nauk. SSSR*, 149 (1963), 245—248. They are influenced, to a certain extent, by earlier papers by Vinogradov, *Izv. Akad. Nauk. SSSR*. Ser. Math. 15 (1951), 109; ibid. 22 (1958), 161; U. V. Linnik, *Mat. Sbornik*, 54 (1943), 28—39; L. K. Hua, *Quarterly J. Math.* (Oxford) 20 (1949), 48; A. A. Karacuba, *Vestnik Moskov. Univ.* Ser. I. Math. Mech. (1962), no. 4, 28; N. M. Korobov, *Doklady Akad. Nauk. SSSR*. 123 (1958), 28.

For a proof of Bertrand's postulate used in Theorem 1, see, for instance, the author's *Introduction*, 71.

For the algebraic preliminaries needed to estimate T_1, see B. L. van der Waerden, *Moderne Algebra*, § 26.

§§ 4—6. For Lemma 2, Theorems 4 and 5, see Vinogradov's book, loc. cit., Ch. VI; also Titchmarsh's *Zeta-function*, 113—114.

For Chudakov's theorem, see *Mat. Sbornik*, (1) 43 (1936), 591—602. The proof given here uses Karacuba's work published in 1963.

Sharper results than Theorem 6 are known. Vinogradov (*Izv. Akad. Nauk. SSSR*, Ser. Math. 22 (1958), 161—164) and Korobov (*Uspehi Mat. Nauk.* 13:4 (1958), 185—192) have proved that $\zeta(s)$ does not vanish in the region $\sigma \geqslant 1 - \frac{A}{(\log t)^\alpha}$, $t \geqslant 3$, $\alpha > \frac{2}{3}$. See Walfisz, loc. cit. Chapters II and V. A sharper result than Theorem 7 can be obtained by using the best-known results on the zero-free region of $\zeta(s)$. Walfisz (loc. cit. Ch. V) gives a proof of the result: $\pi(x) - \operatorname{li} x = O(x e^{-c(\log x)^{3/5}(\log\log x)^{-1/5}})$.

Chapter V

Theorems of Hoheisel and of Ingham

§ 1. The difference between consecutive primes. The prime number theorem implies that $p_n \sim n \log n$, as $n \to \infty$, where p_n denotes the nth prime. A related problem is to determine the size of the difference $p_{n+1} - p_n$. The purpose of this chapter is to prove a theorem of Ingham's which implies, in particular, that

$$p_{n+1} - p_n = O(p_n^{\frac{5}{8}+\varepsilon}), \quad \text{as} \quad n \to \infty,$$

for every $\varepsilon > 0$.

Ingham's starting point is a theorem of Hoheisel, which asserts the existence of an absolute constant $\theta, 0 < \theta < 1$, such that

$$p_{n+1} - p_n = O(p_n^{\theta}), \quad \text{as} \quad n \to \infty. \tag{1}$$

The proof of Hoheisel's result depends on two propositions concerning the non-real zeros of $\zeta(s)$. The first is Littlewood's theorem proved in Chapter III; the second is an estimate of the function $N(\sigma, T)$, which denotes the number of zeros $\rho = \beta + i\gamma$ of $\zeta(s)$ with $\beta \geqslant \sigma$, $0 < \gamma \leqslant T$. This estimate is of the form

$$N(\sigma, T) = O(T^{b(1-\sigma)} \log^B T), \qquad b > 0, \qquad B \geqslant 0, \tag{2}$$

uniformly for $\frac{1}{2} \leqslant \sigma \leqslant 1$, as $T \to \infty$.

Ingham showed that if

$$\zeta(\tfrac{1}{2} + it) = O(t^c), \tag{3}$$

as $t \to \infty$, where c is a positive, absolute constant, then Hoheisel's assumption (2) is satisfied with $b = 2 + 4c$, $B = 5$. Combining it with Chudakov's theorem on the zero-free region of ζ (instead of Littlewood's), he deduced that (3) implies (1) for any fixed θ, such that

$$\frac{1+4c}{2+4c} < \theta < 1. \tag{4}$$

That (3) is actually true with $c = \frac{1}{6} + \varepsilon$, $\varepsilon > 0$, is a result of Hardy and Littlewood, which leads to $\theta = \frac{5}{8} + \varepsilon$. Better values of c have since been found by more elaborate arguments. They lead automatically to better values of θ.

It is easy to see that we may take $c=\frac{1}{4}+\varepsilon$. For $\zeta(1+it)=O(\log t)$, by Theorem 5 of Chapter II, and the functional equation of ζ gives $\zeta(s)=\chi(s)\zeta(1-s)$, where

$$\chi(s) = \frac{\pi^{-(\frac{1}{2}-\frac{1}{2}s)}\Gamma(\frac{1}{2}-\frac{1}{2}s)}{\Gamma(\frac{1}{2}s)\pi^{-\frac{1}{2}s}}.$$

By Stirling's formula for the gamma-function, $(\chi(s))^{-1}=O(t^{\sigma-\frac{1}{2}})$, as $t\to\infty$, in any strip $-\infty<\alpha\leqslant\sigma\leqslant\beta<+\infty$, so that

$$\overline{\zeta(it)}=\zeta(-it)=\frac{\zeta(1+it)}{\chi(1+it)}=O(t^{\frac{1}{2}}\log t).$$

By applying the Phragmén-Lindelöf principle to the function $\dfrac{\zeta(s)}{\log s}$, we get

$$\frac{\zeta(\frac{1}{2}+it)}{\log(\frac{1}{2}+it)}=O(t^{\frac{1}{4}}), \tag{5}$$

or $\zeta(\frac{1}{2}+it)=O(t^{\frac{1}{4}}\log t)$, which proves (3) with $c=\frac{1}{4}+\varepsilon$.

The assertion that (3) is valid for every $c>0$ is known as the *Lindelöf hypothesis*, and is yet to be proved or disproved. It is known that the Riemann hypothesis implies the Lindelöf hypothesis.

§ 2. Landau's formula for the Chebyshev function ψ. The proof of Hoheisel's theorem requires an asymptotic formula for Chebyshev's function ψ, and the proof of that formula requires a number of analytical lemmas.

LEMMA 1. *If* $c>0$, $T>0$, *then*

$$\left|\int_{c-iT}^{c+iT}\frac{y^s}{s}\,ds-2\pi i\right| \leqslant \frac{2}{T}\,\frac{y^c}{\log y}, \quad \text{for} \quad y>1; \tag{6}$$

$$\left|\int_{c-iT}^{c+iT}\frac{y^s}{s}\,ds\right| \leqslant \frac{2}{T}\,\frac{y^c}{-\log y}, \quad \text{for} \quad 0<y<1. \tag{7}$$

PROOF. If $y>1$, then by Cauchy's theorem,

$$\int_{\mathscr{R}}\frac{y^s}{s}\,ds=2\pi i,$$

the integral being taken, in the positive sense, around the rectangle $\mathscr{R}$, whose vertices are $c-Ui$, $c+Vi$, $-X+Vi$, $-X-Ui$, where $U>0$,

$V>0$, $X>0$. If we keep U, V fixed, and let $X\to\infty$, then the integral along the side $(-X+Vi, -X-Ui)$ tends to zero, since

$$\left|\int_{-X-Ui}^{-X+Vi}\frac{y^s}{s}\,ds\right|\leqslant\frac{y^{-X}}{X}\cdot(U+V)<(U+V)\cdot\frac{1}{X},$$

for $y>1$, $X>0$. Hence

$$\int_{c-iU}^{c+iV}\frac{y^s}{s}\,ds=2\pi i-\int_{-\infty-iU}^{c-iU}\frac{y^s}{s}\,ds+\int_{-\infty+iV}^{c+iV}\frac{y^s}{s}\,ds=2\pi i-I_1+I_2,\quad\text{say,}$$

where

$$|I_1|<\int_{-\infty}^{c}\frac{y^\sigma}{U}\,d\sigma=\frac{y^c}{U\log y},$$

and

$$|I_2|<\frac{y^c}{V\log y},$$

since $y^\sigma=e^{\sigma\log y}$, and $\log y>0$. If we take $U=V=T$, we obtain (6). To prove (7), we take a rectangle to the right of the line $\sigma=c$, instead of to the left, and proceed similarly.

LEMMA 2. *If* $1<c<2$, $T>0$, $0<y<2$, *then*

$$\left|\int_{c-iT}^{c+iT}\frac{y^s}{s}\,ds\right|<10\pi. \tag{8}$$

PROOF. If $0<y\leqslant 1$, we apply Cauchy's theorem to the segment $(c-iT, c+iT)$, and to that part of the circle with the origin as centre, and radius $r=(c^2+T^2)^{\frac{1}{2}}$, which lies to the right of the line $\sigma=c$. Then

$$\left|\int_{c-iT}^{c+iT}\frac{y^s}{s}\,ds\right|<\pi r\cdot\frac{y^c}{r}\leqslant\pi.$$

If $y>1$, we apply Cauchy's theorem to the segment $(c-iT, c+iT)$ and to that part of the same circle which lies to the left of the line $\sigma=c$, so that

$$\left|\int_{c-iT}^{c+iT}\frac{y^s}{s}\,ds-2\pi i\right|<2\pi r\cdot\frac{y^c}{r}<8\pi.$$

Lemmas 1 and 2 are required for the proof of the following

LEMMA 3 (LANDAU). *For* $n \geqslant 2$, *let* $|a_n| < A \log n$, *so that the series* $\sum_{n=1}^{\infty} a_n n^{-s}$ *converges for* $\sigma > 1$ *to sum* $f(s)$. *For* $1 < \eta < 2$, *let*

$$(\eta - 1) \sum_{n=1}^{\infty} |a_n| n^{-\eta} \tag{9}$$

be bounded. Then we have, for $1 < \eta < 2$, $T > 0$, $x > 2$,

$$\left| \int_{\eta - iT}^{\eta + iT} \frac{x^s}{s} f(s) ds - 2\pi i \sum_{n \leqslant x} a_n \right| < c_1 \left(\frac{x^\eta}{T(\eta - 1)} + \frac{x \log^2 x}{T} + \log x \right), \tag{10}$$

where c_1 *is independent of* η, T, *and* x.

NOTE. Constants like $c, c_1, A, \ldots$ may have different values at different occurrences.

PROOF. We have, for $1 < \eta < 2$, $T > 0$, $x > 2$,

$$\int_{\eta - iT}^{\eta + iT} \frac{x^s f(s)}{s} ds - 2\pi i \sum_{n \leqslant x} a_n = \int_{\eta - iT}^{\eta + iT} \frac{x^s}{s} \sum_{n=1}^{\infty} a_n n^{-s} ds - 2\pi i \sum_{n \leqslant x - 1} a_n - 2\pi i a_{[x]}$$

$$= \sum_{n \leqslant x - 1} a_n \left(\int_{\eta - iT}^{\eta + iT} \frac{(x/n)^s}{s} ds - 2\pi i \right) +$$

$$+ \sum_{x - 1 < n \leqslant x + 1} a_n \int_{\eta - iT}^{\eta + iT} \frac{(x/n)^s}{s} ds +$$

$$+ \sum_{n > x + 1}^{\infty} a_n \int_{\eta - iT}^{\eta + iT} \frac{(x/n)^s}{s} ds - 2\pi i a_{[x]}.$$

If we use (6) in the first sum, (8) in the second sum, and (7) in the third sum, we have

$$\left| \int_{\eta - iT}^{\eta + iT} \frac{x^s f(s)}{s} ds - 2\pi i \sum_{n \leqslant x} a_n \right| \leqslant \frac{2}{T} \sum_{n \leqslant x - 1} \frac{|a_n| x^\eta}{n^\eta \log(x/n)} +$$

$$+ 10\pi(|a_{[x]}| + |a_{[x]+1}|) + \frac{2}{T} \sum_{n > x + 1}^{\infty} |a_n| \cdot \frac{x^\eta}{n^\eta \log(n/x)} + 2\pi |a_{[x]}|$$

$$< \frac{2x^\eta}{T} \left(\sum_{n \leqslant x - 1} \frac{|a_n|}{n^\eta \log(x/n)} + \sum_{n > x + 1} \frac{|a_n|}{n^\eta \log(n/x)} \right) + c_2 \log x. \tag{11}$$

Now

$$\sum_{n \leqslant x-1} \frac{|a_n|}{n^\eta \log(x/n)} = \sum_{n=1}^{[x/2]} + \sum_{[x/2]+1}^{[x]-1}$$

$$\leqslant \frac{1}{\log 2} \sum_{n=1}^{\infty} \frac{|a_n|}{n^\eta} + \frac{A \log x}{(x/2)^\eta} \sum_{n=[x/2]+1}^{[x]-1} \frac{1}{\log([x]/n)}$$

$$< \frac{c_3}{\eta - 1} + \frac{c_4 \log x}{x^\eta} \sum_{v=1}^{[x]-[x/2]-1} \frac{1}{\log([x]/\{[x]-v\})}$$

by assumption (9). Since $-\log(1-x) > x$, for $0 < x < 1$, we have

$$\sum_{v=1}^{[x]-[\frac{1}{2}x]-1} \frac{1}{\log([x]/\{[x]-v\})} \leqslant \sum_{v=1}^{[x]-1} \frac{1}{(v/[x])}$$

$$\leqslant [x] \sum_{v=1}^{[x-1]} \frac{1}{v} < c x \log x,$$

so that

$$\sum_{n \leqslant x-1} \frac{|a_n|}{n^\eta \log(x/n)} < \frac{c_3}{\eta - 1} + \frac{c_5 \log^2 x}{x^{\eta-1}}. \tag{12}$$

On the other hand,

$$\sum_{n > x+1} \frac{|a_n|}{n^\eta \log(n/x)} = \sum_{n=[x]+2}^{[2x]} + \sum_{[2x]+1}^{\infty}$$

$$< \frac{A \log 2x}{x^\eta} \sum_{n=[x]+2}^{[2x]} \frac{1}{\log\left(\dfrac{n}{[x]+1}\right)} + \frac{1}{\log 2} \sum_{n=1}^{\infty} \frac{|a_n|}{n^\eta}$$

$$< \frac{c_6 \log x}{x^\eta} \sum_{v=1}^{[2x]-[x]-1} \frac{1}{\log\left(\dfrac{[x]+1+v}{[x]+1}\right)} + \frac{c_3}{\eta - 1}$$

$$\leqslant \frac{c_6 \log x}{x^\eta} \sum_{v=1}^{[x]} \frac{1}{\log\left(1 + \dfrac{v}{[x]+1}\right)} + \frac{c_3}{\eta - 1}.$$

Since $\log(1+x)>\frac{1}{2}x$, for $0<x<1$, we have

$$\sum_{n>x+1} \frac{|a_n|}{n^\eta \log(n/x)} \leqslant 2c_6 \frac{\log x}{x^\eta}\cdot([x]+1)\sum_{v=1}^{[x]} \frac{1}{v} + \frac{c_3}{\eta-1}$$
$$< c_7 \frac{\log^2 x}{x^{\eta-1}} + \frac{c_3}{\eta-1}. \tag{13}$$

On using (13) and (12) in (11), we get (10), and hence the lemma.

LEMMA 4. *In the quarter-planes* $\sigma\leqslant -1$, $|t|\geqslant 1$, *as well as on the lines* $\sigma=-u$, *where* u *is an odd integer* $\geqslant 3$, *we have*

$$\left|\frac{\zeta'(s)}{\zeta(s)}\right| < c_1 \log|s|. \tag{14}$$

PROOF. By Stirling's formula, we have for $\frac{1}{2}\leqslant\sigma\leqslant 2$, $t\geqslant 1$, and for $\sigma\geqslant 2$,

$$\left|\frac{\Gamma'(s)}{\Gamma(s)}\right| < c_2 \log|s|. \tag{15}$$

In the half-planes $|t|\geqslant 1$, and on the lines $\sigma=-u$, where u is any odd integer, we have

$$|\cot(\tfrac{1}{2}\pi s)| < c_3. \tag{16}$$

For $t\geqslant 1$ we have

$$|\cot(\tfrac{1}{2}\pi s)| = \left| i\frac{e^{i\pi s}+1}{1-e^{+i\pi s}}\right| \leqslant \frac{e^{-\pi t}+1}{1-e^{-\pi t}} < \frac{2}{1-e^{-\pi}},$$

and similarly also for $t\leqslant -1$. If $-1\leqslant t\leqslant 1$, and $\sigma=-u=-1$, then $\cot(\frac{1}{2}\pi s)$ is regular, and bounded. By periodicity it is so for $-1\leqslant t \leqslant 1$, $\sigma=-u$, where u is any odd integer.

The functional equation of $\zeta(s)$ can be written in the form

$$\zeta(s) = \frac{(2\pi)^s\cdot\sin(\frac{1}{2}\pi s)\cdot\zeta(1-s)}{\sin\pi s\cdot\Gamma(s)} = 2^s\pi^{s-1}\sin(\tfrac{1}{2}\pi s)\cdot\Gamma(1-s)\zeta(1-s),$$

if we use the formulae $\Gamma(s)\Gamma(1-s) = \dfrac{\pi}{\sin\pi s}$, $\Gamma(\frac{1}{2}s)\Gamma(\frac{1}{2}s+\frac{1}{2}) = \dfrac{\sqrt{\pi}\cdot\Gamma(s)}{2^{s-1}}$.

By logarithmic differentiation, we get

$$\frac{\zeta'(s)}{\zeta(s)} = \log 2\pi + \tfrac{1}{2}\pi\cot(\tfrac{1}{2}\pi s) - \frac{\Gamma'(1-s)}{\Gamma(1-s)} - \frac{\zeta'(1-s)}{\zeta(1-s)}.$$

Since the hypotheses of the lemma imply that $1-\sigma \geqslant 2$, $|s| \geqslant \sqrt{2}$, we get, by (15),

$$\left|\frac{\Gamma'(1-s)}{\Gamma(1-s)}\right| < c_2 \log|1-s| \leqslant c_2 \log(1+|s|) < c_4 \log|s|.$$

By (16) we have

$$|\log 2\pi + \tfrac{1}{2}\pi \cot(\tfrac{1}{2}\pi s)| < c_5.$$

Further

$$\left|-\frac{\zeta'(1-s)}{\zeta(1-s)}\right| \leqslant \sum_{n=1}^{\infty} \frac{\Lambda(n)}{n^2} < c_6.$$

Hence

$$\left|\frac{\zeta'(s)}{\zeta(s)}\right| < c_4 \log|s| + c_7 < c_1 \log|s|,$$

which proves (14).

LEMMA 5. *Let $\rho = \beta + i\gamma$ denote any non-real zero of $\zeta(s)$. Then, for $t \geqslant 2$, $t \neq \gamma$, $-1 \leqslant \sigma \leqslant 2$, we have*

$$\left|\frac{\zeta'(s)}{\zeta(s)} - \sum_{|\gamma - t| < 1} \frac{1}{s-\rho}\right| < c \log t. \tag{17}$$

PROOF. We shall first prove that

$$\left|\sum_{|\gamma - t| \geqslant 1} \left(\frac{1}{s-\rho} + \frac{1}{\rho}\right)\right| < c_1 \log t. \tag{18}$$

The infinite product for $\zeta(s)$ gives (§ 3, (46) of Chapter II)

$$\frac{\zeta'(s)}{\zeta(s)} = b - \frac{1}{s-1} - \frac{1}{2}\frac{\Gamma'(\frac{1}{2}s+1)}{\Gamma(\frac{1}{2}s+1)} + \sum_{\rho}\left(\frac{1}{s-\rho} + \frac{1}{\rho}\right), \tag{19}$$

where b is a constant. We replace s by $s+3$, and note that $\sigma+3 \geqslant 2$, and $\dfrac{\sigma}{2} + \dfrac{5}{2} \geqslant 2$. Since $|\zeta'(s)/\zeta(s)| = O(1)$, for $\sigma \geqslant 2$, and since (15) holds, we have

$$\left|\sum_{\rho}\left(\frac{1}{s+3-\rho} + \frac{1}{\rho}\right)\right| < c_2 \log t. \tag{20}$$

A part of this sum is

$$\sum_{|\gamma - t| < 1}\left(\frac{1}{s+3-\rho} + \frac{1}{\rho}\right).$$

It has not more than $N(t+1)-N(t-1)=O(\log t)$ terms, by the Riemann-von Mangoldt formula (§ 2, Ch. II). Each term, in absolute value, is $\leqslant 1+\dfrac{1}{|\rho|}<c_3$, say. Hence

$$\left|\sum_{|\gamma-t|<1}\left(\frac{1}{s+3-\rho}+\frac{1}{\rho}\right)\right|<c_4\log t,$$

and because of (20), we have

$$\left|\sum_{|\gamma-t|\geqslant 1}\left(\frac{1}{s+3-\rho}+\frac{1}{\rho}\right)\right|<c_5\log t. \tag{21}$$

Now

$$\begin{aligned}
&\left|\sum_{\gamma\geqslant t+1}\left(\frac{1}{s-\rho}+\frac{1}{\rho}\right)-\sum_{\gamma\geqslant t+1}\left(\frac{1}{s+3-\rho}+\frac{1}{\rho}\right)\right|\\
&=\left|\sum_{\gamma\geqslant t+1}\left(\frac{1}{s-\rho}-\frac{1}{s+3-\rho}\right)\right|=\left|\sum_{\gamma\geqslant t+1}\frac{3}{(s-\rho)(s+3-\rho)}\right|\\
&\leqslant\sum_{\gamma-t\geqslant 1}\frac{3}{|t-\gamma|\cdot|t-\gamma|}=3\sum_{\gamma\geqslant t+1}\frac{1}{(t-\gamma)^2},
\end{aligned}$$

while

$$3\sum_{n=1}^{\infty}\sum_{t+n\leqslant\gamma<t+n+1}\frac{1}{(t-\gamma)^2}\leqslant 3\sum_{n=1}^{\infty}\sum_{t+n\leqslant\gamma<t+n+1}\frac{1}{n^2}<c_6\sum_{n=1}^{\infty}\frac{\log(t+n)}{n^2},$$

by the Riemann-von Mangoldt formula. Further

$$\sum_{n=1}^{\infty}\frac{\log(t+n)}{n^2}<\sum_{n\leqslant t}\frac{\log 2t}{n^2}+\sum_{n>t}\frac{\log 2n}{n^2}<c_7\log t.$$

Hence

$$\left|\sum_{\gamma\geqslant t+1}\left(\frac{1}{s-\rho}-\frac{1}{s+3-\rho}\right)\right|<c_7\log t,$$

and similarly

$$\left|\sum_{\gamma\leqslant t-1}\left(\frac{1}{s-\rho}-\frac{1}{s+3-\rho}\right)\right|<c_8\log t,$$

so that, taken together with (21), they give (18). The lemma now follows from (19), (18) and (15) and the Riemann-von Mangoldt formula.

We are now in a position to prove the following

THEOREM 1 (LANDAU). *Let* $\rho=\beta+i\gamma$ *denote any non-real zero of* $\zeta(s)$. *Then there exists an absolute constant* c, *such that for* $x\geqslant 3$, $T\geqslant 3$, *we have*

$$\left|\psi(x)-x+\sum_{|\gamma|<T}\frac{x^\rho}{\rho}\right|\leqslant c\left(\frac{x\log^2 x}{T}+\frac{x\log T}{T}+\log x\right). \tag{22}$$

PROOF. Let $x>1; 2>\eta>1$; $T>2$; $T\neq\gamma$ for any zero ρ; u an odd integer $\geqslant 3$. We apply Cauchy's theorem to the integral

$$\int_{\mathscr{R}}\frac{x^s}{s}\frac{\zeta'(s)}{\zeta(s)}\,ds,$$

taken in the positive sense along the rectangle $\mathscr{R}$ with vertices at $\eta\pm iT$, $-u\pm iT$. We then have

$$\frac{1}{2\pi i}\int_{\mathscr{R}}\frac{x^s}{s}\frac{\zeta'(s)}{\zeta(s)}\,ds=-x+\sum_{|\gamma|<T}\frac{x^\rho}{\rho}+\sum_{n=1}^{\frac{1}{2}(u-1)}\frac{x^{-2n}}{-2n}+\frac{\zeta'(0)}{\zeta(0)}.$$

We now let $u\to\infty$. Then the integral along the side $(-u+iT, -u-iT)$ tends to zero, since, for s lying on that segment, we have

$$\left|\frac{1}{s}\frac{\zeta'(s)}{\zeta(s)}\right|<c_1\frac{\log|s|}{|s|},$$

by (14), and $\dfrac{\log y}{y}$ is monotone decreasing for $y\geqslant e$, while $|s|\geqslant u\geqslant 3>e$, so that

$$\left|\frac{x^s}{s}\frac{\zeta'(s)}{\zeta(s)}\right|<c_1x^{-u}\frac{\log u}{u}<c_1\frac{\log u}{u},$$

and

$$\left|\int_{-u-iT}^{-u+iT}\frac{x^s}{s}\frac{\zeta'(s)}{\zeta(s)}\,ds\right|<2c_1\cdot T\frac{\log u}{u}.$$

For fixed x and T, the integral tends to zero as $u\to\infty$.

Hence we have

$$\frac{1}{2\pi i}\int_{\eta-iT}^{\eta+iT}\frac{x^s}{s}\frac{\zeta'(s)}{\zeta(s)}\,ds=-x+\sum_{|\gamma|<T}\frac{x^\rho}{\rho}+\sum_{n=1}^{\infty}\frac{x^{-2n}}{-2n}+\frac{\zeta'(0)}{\zeta(0)}-$$

$$-\frac{1}{2\pi i}\int_{-\infty-iT}^{\eta-iT}\frac{x^s}{s}\frac{\zeta'(s)}{\zeta(s)}\,ds+\frac{1}{2\pi i}\int_{-\infty+iT}^{\eta+iT}\frac{x^s}{s}\frac{\zeta'(s)}{\zeta(s)}\,ds, \tag{23}$$

the integrals along the sides of the rectangle $\mathscr{R}$ parallel to the real axis being convergent in virtue of (14), because

$$\left|\frac{x^s}{s}\frac{\zeta'(s)}{\zeta(s)}\right| < c_1 x^\sigma \frac{\log|s|}{|s|} < cx^\sigma, \quad \text{for} \quad \sigma \leqslant -1, \ t > 2,$$

and $\int\limits_{-\infty}^{-1} x^\sigma d\sigma$ is convergent.

We shall now estimate the integral

$$\int\limits_{-\infty+iT}^{\eta+iT} \frac{x^s}{s}\frac{\zeta'(s)}{\zeta(s)} ds = \int\limits_{-\infty+iT}^{-1+iT} + \int\limits_{-1+iT}^{\eta+iT},$$

which occurs on the right-hand side of (23). First

$$\left|\int\limits_{-\infty+iT}^{-1+iT} \frac{x^s}{s}\frac{\zeta'(s)}{\zeta(s)} ds\right| < \frac{c_2 \log T}{T}\int\limits_{-\infty}^{-1} x^\sigma d\sigma = \frac{c_2 \log T}{Tx\log x}, \tag{24}$$

because, by (14),

$$\left|\frac{1}{s}\frac{\zeta'(s)}{\zeta(s)}\right| < \frac{c_1 \log|s|}{|s|} \leqslant \frac{c_2 \log T}{T},$$

if we note that $\frac{\log t}{t}$ is decreasing for $t \geqslant e$, and for $2 \leqslant t \leqslant e$, $\frac{\log|s|}{|s|} = O(1)$, and $\frac{t}{\log t} = O(1)$.

Secondly

$$\int\limits_{-1+iT}^{\eta+iT} \frac{x^s}{s}\frac{\zeta'(s)}{\zeta(s)} ds = \int\limits_{-1+iT}^{\eta+iT} \frac{x^s}{s}\left(\frac{\zeta'(s)}{\zeta(s)} - \sum_{|\gamma-T|<1} \frac{1}{s-\rho}\right) ds +$$

$$+ \int\limits_{-1+iT}^{\eta+iT} \sum_{|\gamma-T|<1} \frac{x^s}{s(s-\rho)} ds. \tag{25}$$

By Lemma 5, the first integral on the right-hand side of (25) is, in absolute value,

$$\leqslant \frac{c\log T}{T}\int\limits_{-1}^{\eta} x^\sigma d\sigma < \frac{c\log T}{T}\int\limits_{-\infty}^{\eta} x^\sigma d\sigma = \frac{c\log T}{T}\cdot\frac{x^\eta}{\log x}. \tag{26}$$

The second integral on the right-hand side of (25) is, in absolute value,

$$\leqslant \sum_{|\gamma-T|<1} \left| \int_{-1+iT}^{\eta+iT} \frac{x^s\,ds}{s(s-\rho)} \right|. \tag{27}$$

We deform the path of integration here according as $T-1<\gamma<T$, or $T<\gamma<T+1$. In the former case, we consider a semi-circle C with centre at $\beta+iT$ and radius $\eta-1$ (where $\rho=\beta+i\gamma$ is the non-real zero of $\zeta(s)$ in question) *above* the line $t=T$, and take the path of integration to be the line segment $(-1+iT, \beta-(\eta-1)+iT)$, the semi-circle C above the line $t=T$, and the line segment $(\beta+(\eta-1)+iT, \eta+iT)$. If $T<\gamma<T+1$, however, we take a similar semi-circle C' *below* the line $t=T$. For s on either semi-circle C or C', we have $|s-\rho|>\eta-1$, so that

$$\int_{C \text{ or } C'} \frac{x^s}{s(s-\rho)}\,ds < \pi(\eta-1)\frac{x^{\beta+(\eta-1)}}{T-1} \cdot \frac{1}{\eta-1} \leqslant \pi \cdot \frac{x^{1+\eta-1}}{T-1} < \frac{2\pi x^\eta}{T}, \tag{28}$$

while

$$\left| \int_{-1+iT}^{\beta-(\eta-1)+iT} + \int_{\beta+(\eta-1)+iT}^{\eta+iT} \right| \leqslant \frac{1}{T(\eta-1)} \int_{-1}^{\eta} x^\sigma\,d\sigma \leqslant \frac{x^\eta}{T(\eta-1)\log x}. \tag{29}$$

Because of the Riemann-von Mangoldt formula, the number of terms in (27) is $O(\log T)$. Hence, by (25), (26), (27), (28) and (29), we have

$$\left| \int_{-1+iT}^{\eta+iT} \frac{x^s}{s}\frac{\zeta'(s)}{\zeta(s)}\,ds \right| < c_1 \log T \left(\frac{2\pi x^\eta}{T} + \frac{x^\eta}{T(\eta-1)\log x} \right) + \frac{c \log T}{T} \cdot \frac{x^\eta}{\log x}$$

$$< c_2 \frac{x^\eta \log T}{T} \left(1 + \frac{1}{(\eta-1)\log x} \right).$$

Taken together with (24), this leads to the estimate

$$\left| \int_{-\infty+iT}^{\eta+iT} \frac{x^s}{s}\frac{\zeta'(s)}{\zeta(s)}\,ds \right| < c_3 \frac{x^\eta \log T}{T} \left(1 + \frac{1}{(\eta-1)\log x} \right), \tag{30}$$

for $T>2$, $T \neq \gamma$, $x>1$, $1<\eta<2$.

By symmetry, a similar estimate holds for $\int\limits_{-\infty-iT}^{\eta-iT}$. Hence for $T>2$, $T\neq\gamma$, $x>1$, $1<\eta<2$, we obtain, from (23),

$$\left|\frac{1}{2\pi i}\int\limits_{\eta-iT}^{\eta+iT}\frac{x^s}{s}\frac{\zeta'(s)}{\zeta(s)}ds+x-\frac{1}{2}\log\left(1-\frac{1}{x^2}\right)-\frac{\zeta'(0)}{\zeta(0)}-\sum_{|\gamma|<T}\frac{x^\rho}{\rho}\right|$$
$$<c_4\frac{x^\eta\log T}{T}\left(1+\frac{1}{(\eta-1)\log x}\right). \tag{31}$$

We can now apply Lemma 3 to the function

$$f(s)=\frac{\zeta'(s)}{\zeta(s)}=-\sum_{n=1}^{\infty}\frac{\Lambda(n)}{n^s}=-\sum_{p,m}\frac{\log p}{p^{ms}},$$

because

$$\lim_{\eta\to1+0}(\eta-1)\sum_{p,m}\frac{\log p}{p^{m\eta}}=1,$$

and obtain for $T>0$, $x>2$, $1<\eta<2$,

$$\left|\frac{1}{2\pi i}\int\limits_{\eta-iT}^{\eta+iT}\frac{x^s}{s}\frac{\zeta'(s)}{\zeta(s)}ds+\psi(x)\right|<c_5\left(\frac{x^\eta}{T(\eta-1)}+\frac{x\log^2x}{T}+\log x\right). \tag{32}$$

From (32) and (31) we get for $x>2$, $1<\eta<2$, $T>2$, $T\neq\gamma$,

$$\left|\psi(x)-x+\sum_{|\gamma|<T}\frac{x^\rho}{\rho}\right|$$
$$<c_6\left(\frac{x^\eta\log T}{T}+\frac{x^\eta}{(\eta-1)\log x}\cdot\frac{\log T}{T}+\frac{x^\eta}{\eta-1}\cdot\frac{1}{T}+\frac{x\log^2x}{T}+\log x\right).$$

The left-hand side is independent of η. We now choose η so as to obtain the best inequality here. Let $x\geqslant3$, and $\eta=1+\frac{1}{\log x}$, so that $1<\eta<2$. Then we have for $x\geqslant3$, $T>2$, $T\neq\gamma$,

$$\left|\psi(x)-x+\sum_{|\gamma|<T}\frac{x^\rho}{\rho}\right|<c_7\left(\frac{x\log^2x}{T}+\frac{x\log T}{T}+\log x\right). \tag{33}$$

We can now remove the restriction that T is not the ordinate of a zero, for if it is, we have only to observe that

$$\sum_{|\gamma|<T} \frac{x^\rho}{\rho} = \lim_{\delta\to 0+} \sum_{|\gamma|<T-\delta} \frac{x^\rho}{\rho},$$

so that inequality (33) holds all the same, with $\leqslant$ in place of $<$, and hence the theorem.

§ 3. Hoheisel's theorem. We shall state and prove Hoheisel's theorem in the form given to it by Ingham.

THEOREM 2 (HOHEISEL). *Suppose that*

$$\zeta(s)\neq 0, \quad \textit{for} \quad \sigma>1-\frac{A\log\log t}{\log t}, \qquad t>t_0, \qquad A>0, \qquad t_0>3, \tag{34}$$

and that

$$N(\sigma,T)=O\left(T^{b(1-\sigma)}\log^B T\right), \qquad b>0, \qquad B\geqslant 0, \tag{35}$$

uniformly for $\frac{1}{2}\leqslant\sigma\leqslant 1$, *as* $T\to\infty$. *Then*

$$\pi(x+x^\theta)-\pi(x) \sim \frac{x^\theta}{\log x}, \tag{36}$$

and therefore

$$p_{n+1}-p_n=O\left(p_n^\theta\right), \tag{37}$$

for any fixed θ, *such that*

$$1-\frac{1}{b+B/A}<\theta<1. \tag{38}$$

PROOF. By Theorem 1, we have

$$\psi(x)=x-\sum_{|\gamma|<T}\frac{x^\rho}{\rho}+O\left(\frac{x\log^2 x}{T}\right),$$

uniformly for $3\leqslant T\leqslant x$, as $x\to\infty$, where $\rho=\beta+i\gamma$ is any non-real zero of $\zeta(s)$. Hence

$$\psi(x+h)-\psi(x)=h-\sum_{|\gamma|<T}\frac{(x+h)^\rho-x^\rho}{\rho}+O\left(\frac{x\log^2 x}{T}\right), \tag{39}$$

where the O is uniform for $3\leqslant T\leqslant x$, $0<h\leqslant x$, and $x\to\infty$. Since

$$\left|\frac{(x+h)^\rho-x^\rho}{\rho}\right| = \left|\int_x^{x+h} v^{\rho-1}\,dv\right| \leqslant \int_x^{x+h} v^{\beta-1}\,dv\leqslant h x^{\beta-1},$$

we have

$$\frac{\psi(x+h)-\psi(x)}{h} = 1 + O\left(\sum_{|\gamma|<T} x^{\beta-1}\right) + O\left(\frac{x}{Th}\log^2 x\right). \tag{40}$$

Since

$$\sum_{|\gamma|\leqslant T} (x^{\beta-1} - x^{-1}) = \sum_{|\gamma|\leqslant T} \int_0^\beta x^{\sigma-1}\log x\, d\sigma$$
$$= \int_0^1 \sum_{\substack{|\gamma|\leqslant T\\ \beta\geqslant\sigma}} x^{\sigma-1}\log x\, d\sigma,$$

we have

$$\sum_{|\gamma|\leqslant T} x^{\beta-1} = 2x^{-1}N(0,T) + 2\int_0^1 N(\sigma,T)x^{\sigma-1}\cdot\log x\cdot d\sigma. \tag{41}$$

By the Riemann-von Mangoldt formula, $N(0,T)=O(T\log T)$. By hypothesis (34), there exists a $T_0>3$, such that $N(\sigma,T)=0$ for $T\geqslant T_0$, and $\sigma>1-\eta(T)$, $\eta(T) = \dfrac{A\log\log T}{\log T}$, $A>0$. Further $N(\sigma,T)\leqslant 2N(\frac{1}{2},T)$ for $\sigma\leqslant\frac{1}{2}$, so that (35) holds uniformly for $0\leqslant\sigma\leqslant 1$. Hence (41) leads to

$$\sum_{|\gamma|\leqslant T} x^{\beta-1} = O\left(\frac{T\log T}{x}\right) + O\left(\int_0^{1-\eta(T)} \left(\frac{T^b}{x}\right)^{1-\sigma}\cdot\log^B T\cdot\log x\cdot d\sigma\right),$$

uniformly for $T_0\leqslant T\leqslant x$, as $x\to\infty$.

We take $T=x^\alpha$, where $0<\alpha<\dfrac{1}{b}\leqslant\dfrac{1}{2}$ (by (35) and the Riemann-von Mangoldt formula). Then

$$\sum_{|\gamma|\leqslant T} x^{\beta-1} = O(x^{\alpha-1}\log x) + O(x^{(\alpha b-1)\eta(x^\alpha)}\log^B x)$$
$$= O(x^{\alpha-1}\log x) + O\left(e^{\left(b-\frac{1}{\alpha}\right)A\log(\alpha\log x)}\log^B x\right)$$
$$= O((\log x)^{-\delta}), \tag{42}$$

where $\delta = \left(\dfrac{1}{\alpha} - b\right)A - B$.

If we now choose α, such that

$$\frac{1}{\alpha} > b + \frac{B}{A} \quad (\geqslant b, \text{ since } B\geqslant 0,\ A>0),$$

then $\delta > 0$, so that by (42) and (40) we have

$$\psi(x+h)-\psi(x) \sim h, \quad \text{as} \quad x \to \infty, \tag{43}$$

if $h = x^\theta$, and θ is a constant such that $1 > \theta > 1-\alpha(>\frac{1}{2})$.

If $h = x^\theta$, however, we have

$$\begin{aligned}\psi(x+h)-\psi(x) &= \sum_{x<p\leqslant x+h} \log p + O\left(\sum_{p^2\leqslant x+h} \log p \cdot \left[\frac{\log(x+h)}{\log p}\right]\right) \\ &= \sum_{x<p\leqslant x+h} (\log x + O(1)) + O\left(\sum_{p^2 \leqslant 2x} \log 2x\right) \\ &= (\pi(x+h)-\pi(x))(\log x + O(1)) + O(x^{\frac{1}{2}} \log x).\end{aligned}$$

This together with (43) gives us (36), if we note the conditions on α and θ; for any given θ satisfying (38), an appropriate α can be found. If we take $x = p_n$ in (36), we get (37).

REMARK. Assumption (34) of Theorem 2 is satisfied because of Littlewood's theorem (Ch. III).

§ 4. Two auxiliary lemmas. We need for subsequent use a convexity theorem for integrals due to Hardy, Ingham, and Pólya, as well as some arithmetical estimates, involving the divisor function, due to Ingham.

LEMMA 6 (HARDY, INGHAM, PÓLYA). *If $f(s)$ is regular and bounded for $\sigma_1 \leqslant \sigma \leqslant \sigma_2$, and the integral*

$$J(\sigma) = \int_{-\infty}^{\infty} |f(\sigma+it)|^2 \, dt \tag{44}$$

exists, and converges uniformly for $\sigma_1 \leqslant \sigma \leqslant \sigma_2$, and

$$\lim_{|t|\to\infty} |f(s)| = 0,$$

uniformly for $\sigma_1 \leqslant \sigma \leqslant \sigma_2$, then

$$J(\sigma) \leqslant \{J(\sigma_1)\}^{\frac{\sigma_2-\sigma}{\sigma_2-\sigma_1}} \{J(\sigma_2)\}^{\frac{\sigma-\sigma_1}{\sigma_2-\sigma_1}}. \tag{45}$$

PROOF. Initially we prove this result for $\sigma = \sigma_0 = \frac{1}{2}(\sigma_1+\sigma_2)$. By hypothesis, $f(s)$ is regular in a rectangle $\mathcal{R}$ of the form

$$\sigma_1 \leqslant \sigma \leqslant \sigma_2, \qquad -T \leqslant t \leqslant T, \qquad \sigma_1 < \sigma_2.$$

So also is the function $f^*(s) = \overline{f(2\sigma_0 - \bar{s})}$, where the bar denotes the complex conjugate. Since $f^*(s) = \overline{f(s)}$ for $\operatorname{Re} s = \sigma_0$, we have

$$i \int_{\sigma_0-iT}^{\sigma_0+iT} |f(s)|^2 \, dt = \int_{\sigma_0-iT}^{\sigma_0+iT} f(s) f^*(s) \, ds = \int_{\mathcal{R}_2} f(s) f^*(s) \, ds,$$

where $\mathscr{R}_2$ denotes that part of the boundary of the rectangle $\mathscr{R}$, which lies to the right of the line $\operatorname{Re}s=\sigma_0$, taken in the positive sense. However,

$$\begin{aligned}\left|\int_{\mathscr{R}_2} f(s)f^*(s)\,ds\right| &\leqslant \int_{\mathscr{R}_2} |f(s)|\cdot|f^*(s)|\cdot|ds| \\ &\leqslant \Big(\int_{\mathscr{R}_2} |f(s)|^2\,|ds|\cdot\int_{\mathscr{R}_2} |f^*(s)|^2\,|ds|\Big)^{\frac{1}{2}} \\ &= \Big(\int_{\mathscr{R}_2} |f(s)|^2\,|ds|\Big)^{\frac{1}{2}}\Big(\int_{\mathscr{R}_1} |f(s)|^2\,|ds|\Big)^{\frac{1}{2}}, \end{aligned}\tag{46}$$

where $\mathscr{R}_1$ is that part of the boundary of $\mathscr{R}$ which lies to the left of the line $\operatorname{Re}s=\sigma_0$, taken in the negative sense, since $2\sigma_0-\bar{s}\in\mathscr{R}_1$ whenever $s\in\mathscr{R}_2$.

Since $|f(s)|\to 0$, as $|t|\to\infty$, by hypothesis, we have

$$\lim_{T\to\pm\infty}\int_{\sigma_0+iT}^{\sigma_2+iT} |f(s)|^2\,|ds|=0,$$

which implies that

$$\lim_{T\to\infty}\int_{\mathscr{R}_2} |f(s)|^2\,|ds| = \int_{-\infty}^{\infty} |f(\sigma_2+it)|^2\,dt=J(\sigma_2),\tag{47}$$

and similarly

$$\lim_{T\to\infty}\int_{\mathscr{R}_1} |f(s)|^2\cdot|ds|=J(\sigma_1).\tag{48}$$

If we use (48) and (47) in (46), we see that (45) holds for $\sigma=\frac{1}{2}(\sigma_1+\sigma_2)$.

We next show that if (45) holds for $\sigma=\sigma_1'$ and $\sigma=\sigma_2'$, then it holds also for $\sigma=\sigma_0'=\frac{1}{2}(\sigma_1'+\sigma_2')$. This is immediate if we just write down the inequalities given by (45) for $\sigma=\sigma_1'$, and $\sigma=\sigma_2'$, and use them in the inequality we have just established, namely

$$J(\sigma_0')\leqslant J(\sigma_1')^{\frac{\sigma_2'-\sigma_0'}{\sigma_2'-\sigma_1'}}\cdot J(\sigma_2')^{\frac{\sigma_0'-\sigma_1'}{\sigma_2'-\sigma_1'}}.$$

Thus we have

$$J(\sigma_0')\leqslant J(\sigma_1)^{\frac{\sigma_2-\sigma_0'}{\sigma_2-\sigma_1}}\cdot J(\sigma_2)^{\frac{\sigma_0'-\sigma_1}{\sigma_2-\sigma_1}}.$$

But inequality (45) is trivially true for $\sigma=\sigma_1$, and for $\sigma=\sigma_2$, so it holds for all σ of the form $\sigma=\sigma_1+(\sigma_2-\sigma_1)\dfrac{m}{2^n}$, $0\leqslant m\leqslant 2^n$, $n=1,2,\ldots$. Since $J(\sigma)$ is a continuous function of σ by hypothesis, it holds for all σ, such that $\sigma_1\leqslant\sigma\leqslant\sigma_2$.

Lemma 7. *Let $d(n)$ denote the number of positive divisors of the positive integer n. Then*

$$\sum_{n\leqslant x} d^2(n) < c_1 x\log^3 x, \qquad x\geqslant 2, \tag{49}$$

and

$$\sum\sum_{m<n\leqslant x} \frac{d(m)d(n)}{(mn)^{\frac{1}{2}}\log(n/m)} < c_2 x\log^3 x, \qquad x>1, \tag{50}$$

where c_1 and c_2 are positive, absolute constants.

Proof. It is an elementary fact that

$$\sum_{n\leqslant x} d(n) = O(x\log x).$$

(see Notes on Ch. I). Now

$$\sum_{n\leqslant x} d^2(n) = \sum_{n\leqslant x} d(n)\sum_{k|n} 1 = \sum_{k\leqslant x}\sum_{km=n\leqslant x} d(n) \leqslant \sum_{k\leqslant x}\sum_{m\leqslant\frac{x}{k}} d(k)d(m)$$

$$= \sum_{k\leqslant x} d(k)\sum_{m\leqslant\frac{x}{k}} d(m) = O\left(x\sum_{k\leqslant x}\frac{d(k)}{k}\log\left(\frac{x}{k}+1\right)\right) = O(x\log^3 x),$$

by partial summation, which proves (49).

To prove (50), we first prove the following proposition: if k is a positive integer, and $x\geqslant k$, then

$$\sum = \sum_{1\leqslant r\leqslant x-k} d(r)d(r+k) < 4x(\log x+1)^2\sum_{d|k}\frac{1}{d}. \tag{51}$$

The sum on the left-hand side can be written as

$$\sum = \sum_{\substack{\mu m-\nu n=k\\ \mu m\leqslant x}} 1,$$

the summation in each of the variables μ, ν, m, n, starting from 1. If S denotes the sum of all those terms for which $mn\leqslant x$, and S' the sum of all those terms for which $\mu\nu\leqslant x$, then each term of $\sum$ belongs to one at least of these two sums, since

$$\mu\nu\cdot mn = \mu m\cdot\nu n < (\mu m)^2 \leqslant x^2.$$

Further $S=S'$ by symmetry. Hence

$$\sum \leqslant S+S' = 2S = 2\sum_{mn\leqslant x} N(m,n),$$

where $N(m,n)=N(m,n,x,k)$ is the number of pairs of integers μ, ν satisfying the conditions

$$\mu m - \nu n = k, \qquad 0<\mu \leqslant \frac{x}{m}, \qquad \nu > 0.$$

Let $(m,n)=d$, $m=dm'$, $n=dn'$. If d does not divide k, clearly $N(m,n)=0$. If d divides k, the equation $\mu m - \nu n = k$ has integral solutions, and these are all given by the formula

$$\mu = \mu_0 + \frac{rn}{d}, \qquad \nu = \nu_0 + \frac{rm}{d},$$

where μ_0, ν_0 is a particular solution, and r an arbitrary integer. Thus $N(m,n)$ is at most equal to the number of integers r satisfying

$$0 < \mu_0 + \frac{rn}{d} \leqslant \frac{x}{m};$$

that is,

$$N(m,n) \leqslant \left[\frac{xd}{mn} - \frac{\mu_0 d}{n}\right] - \left[-\frac{\mu_0 d}{n}\right] < \frac{xd}{mn} + 1 \leqslant \frac{2xd}{mn} = \frac{2x}{dm'n'},$$

if $mn \leqslant x$. Hence

$$\sum < 4x \sum_{d|k} \frac{1}{d} \sum_{\substack{m'n' \leqslant \frac{x}{d^2} \\ (m',n')=1}} \frac{1}{m'n'} \leqslant 4x \sum_{d|k} \frac{1}{d} \Bigg(\sum_{1 \leqslant m' \leqslant x} \frac{1}{m'} \Bigg)^2$$
$$\leqslant 4x \sum_{d|k} \frac{1}{d} (\log x + 1)^2,$$

which proves (51).

Now, if $1 \leqslant m < n$, we have

$$\left|\log\left(\frac{m}{n}\right)\right|^{-1} = \left\{-\log\left(1 - \frac{n-m}{n}\right)\right\}^{-1} < \frac{n}{n-m}$$
$$= 1 + \frac{m}{n-m} < 1 + \frac{(mn)^{\frac{1}{2}}}{n-m}.$$

Hence

$$\sum_{1 \leqslant m < n \leqslant x}\sum \frac{d(m)d(n)}{(mn)^{\frac{1}{2}} \left|\log \frac{m}{n}\right|} < \sum_{1 \leqslant m < n \leqslant x}\sum \frac{d(m)d(n)}{(mn)^{\frac{1}{2}}} + \sum_{1 \leqslant m < n \leqslant x}\sum \frac{d(m)d(n)}{n-m}. \quad (52)$$

Now

$$\sum_{1 \leqslant m < n \leqslant x} \sum \frac{d(m)d(n)}{(mn)^{\frac{1}{2}}} < \left(\sum_{1}^{[x]} \frac{d(n)}{\sqrt{n}} \right)^2 = O(\sqrt{x} \log x)^2, \tag{53}$$

while, on writing $n - m = k$, we have

$$\begin{aligned} \sum_{1 \leqslant m < n \leqslant x} \sum \frac{d(m)d(n)}{n-m} &= \sum_{k=1}^{[x-1]} \frac{1}{k} \sum_{m=1}^{[x-k]} d(m)d(m+k) \\ &= O(x \log^2 x) \sum_{dd' \leqslant x} \frac{1}{dd'} \cdot \frac{1}{d} \quad \text{(by 51)} \\ &= O(x \log^2 x) \sum_{d=1}^{\infty} \frac{1}{d^2} \sum_{d'=1}^{x} \frac{1}{d'} \\ &= O(x \log^3 x). \end{aligned} \tag{54}$$

If we use (54) and (53) in (52), we get (50).

§ 5. Ingham's theorem. In order to prove that an estimate of the form $\zeta(\frac{1}{2} + it) = O(t^c)$ implies an estimate for $N(\sigma, T)$ of the form given by (35), we need the following

LEMMA 8 (INGHAM). *Let*

$$f_X(s) = \zeta(s) \sum_{n < X} \mu(n) n^{-s} - 1 = \zeta(s) M_X(s) - 1, \tag{55}$$

where $\mu(n)$ *denotes the Möbius function. If*

$$\zeta(\tfrac{1}{2} + it) = O(t^c), \quad \textit{as} \quad t \to \infty, \tag{3}$$

where c *denotes a positive, absolute constant, then*

$$\int_1^T |f_X(\sigma + it)|^2 \, dt < A \frac{T^{4c(1-\sigma)}}{X^{2\sigma - 1}} \cdot (T + X) \log^4(T + X), \tag{56}$$

for $\frac{1}{2} \leqslant \sigma \leqslant 1$, $T > 1$, $X > 1$, *where* A *is a positive, absolute constant.*

PROOF. For $1 < X < 2$, we have $f_X(s) = f_2(s)$, so that we may suppose, in what follows, that $X \geqslant 2$.

For $\sigma > 1$, we have

$$f_X(s) = \zeta(s) \sum_{n < X} \mu(n) n^{-s} - 1 = \sum_{n=1}^{\infty} \frac{a_X(n)}{n^s},$$

where $a_X(1)=0$, and for $n>1$,

$$a_X(n) = \sum_{\substack{d|n \\ d<X}} \mu(d), \tag{57}$$

so that $a_X(n)=0$ if $1<n<X$, and $|a_X(n)|\leqslant d(n)$, for all n and X.

We seek to estimate $\int_0^T |f_X(\sigma+it)|^2\,dt$ for $\sigma=1+\delta, 0<\delta<1$, and for $\sigma=\frac{1}{2}$, and then, by Lemma 6, obtain an estimate valid for $\frac{1}{2}\leqslant\sigma\leqslant 1+\delta$.

If $0<\delta<1$, we have for $T>0$,

$$\begin{aligned}\int_0^T |f_X(1+\delta+it)|^2\,dt &= \sum_{m,n\geqslant X}\sum \frac{a_X(m)a_X(n)}{(mn)^{1+\delta}} \int_0^T \left(\frac{m}{n}\right)^{it} dt \\ &= \sum_{m=n\geqslant X}\sum + 2\,\mathrm{Re} \sum_{X\leqslant m<n}\sum \\ &\leqslant T \sum_{n\geqslant X} \frac{d^2(n)}{n^{2+2\delta}} + 4 \sum_{n>m\geqslant X}\sum \frac{d(m)d(n)}{(mn)^{1+\delta}\log(n/m)}.\end{aligned} \tag{58}$$

We shall estimate the last two sums with the help of Lemma 7. If $0<\xi<3$, then

$$\begin{aligned}\sum_{n\geqslant X} \frac{d^2(n)}{n^{1+\xi}} &= \sum_{n\geqslant X} d^2(n) \int_n^\infty \frac{1+\xi}{x^{2+\xi}}\,dx \\ &= \int_X^\infty \frac{1+\xi}{x^{2+\xi}} \Bigl(\sum_{X\leqslant n\leqslant x} d^2(n)\Bigr) dx \\ &< \int_X^\infty \frac{(1+\xi)c_1\log^3 x}{x^{1+\xi}}\,dx \quad \text{(by (49))} \\ &< \frac{c_3}{\xi X^\xi}\left(\frac{1}{\xi} + \log X\right)^3,\end{aligned} \tag{59}$$

by the substitution $x=Xy^{1/\xi}$.

To estimate the second sum in (58), we observe that for $\lambda > 1$, we have

$$1 < \log\lambda + \frac{1}{\lambda} < \log\lambda + \frac{1}{\sqrt{\lambda}},$$

so that

$$\sum_{n>m\geqslant X}\sum \frac{d(m)d(n)}{(mn)^{1+\xi}\log(n/m)} < \sum_{n>m\geqslant X}\sum \frac{d(m)d(n)}{(mn)^{1+\xi}} + \sum_{n>m\geqslant X}\sum \frac{d(m)d(n)}{m^{\xi}n^{1+\xi}(mn)^{\frac{1}{2}}\log(n/m)}$$

$$< \left(\sum_{n=1}^{\infty} \frac{d(n)}{n^{1+\xi}}\right)^2 + \sum_{n>m\geqslant 1}\sum \frac{d(m)d(n)}{(mn)^{\frac{1}{2}}\log(n/m)} \int_n^{\infty} \frac{1+\xi}{x^{2+\xi}}\,dx$$

$$= \zeta^4(1+\xi) + \int_1^{\infty} \frac{1+\xi}{x^{2+\xi}}\left(\sum_{m<n\leqslant x}\sum \frac{d(m)d(n)}{(mn)^{\frac{1}{2}}\log(n/m)}\right)dx$$

$$< \zeta^4(1+\xi) + \int_1^{\infty} \frac{1+\xi}{x^{1+\xi}}\cdot c_2 \log^3 x\,dx \quad \text{(by (50))}$$

$$< \frac{c_4}{\xi^4}. \tag{60}$$

We use (59) with $\xi = 1+2\delta$, and (60) with $\xi = \delta$, in (58), and obtain

$$\int_0^T |f_X(1+\delta+it)|^2\,dt < c_5\left(\frac{T}{X}+1\right)\delta^{-4}, \tag{61}$$

for (59) gives

$$\sum_{n\geqslant X} \frac{d^2(n)}{n^{2+2\delta}} < c_6 \frac{\log^3 X}{X^{1+2\delta}} < \frac{c_7}{X\delta^3},$$

since $X^{2\delta} = e^{2\delta\log X} > \frac{1}{6}(2\delta\log X)^3$.

For $\sigma = \frac{1}{2}$, we use the inequalities

$$|f_X|^2 \leqslant 2(|\zeta|^2|M_X|^2 + 1),$$

and

$$\int_0^T |M_X(\tfrac{1}{2}+it)|^2\,dt \leqslant T\sum_{n<X} \frac{\mu^2(n)}{n} + 4\sum_{m<n<X}\sum \frac{|\mu(m)\cdot\mu(n)|}{(mn)^{\frac{1}{2}}\log(n/m)}$$

$$\leqslant T\sum_{n<X}\frac{1}{n} + 4\sum_{m<n<X}\sum\left(\frac{1}{(mn)^{\frac{1}{2}}} + \frac{1}{n-m}\right)$$

$$< c_8(T+X)\log X,$$

since $\dfrac{1}{\log\lambda} < \dfrac{\lambda}{\lambda-1} < 1 + \dfrac{\lambda^{\frac{1}{2}}}{\lambda-1}$, for $\lambda > 1$. If we use (3), we get for $T > 1$,

$$\int_0^T |f_X(\tfrac{1}{2}+it)|^2\,dt \leqslant c_9 \cdot T^{2c}(T+X)\log X, \tag{62}$$

the inequality holding down to $T=0$, since the integral is at most $c_{10} \cdot X \cdot T \leqslant c_{10} \cdot T^{2c} X$ for $0 \leqslant T \leqslant 1$ (since $0 < c < \frac{1}{2}$, § 1).

From (61) and (62) we shall deduce an inequality valid for $\frac{1}{2} \leqslant \sigma \leqslant 1+\delta$ by means of Lemma 6.

Let

$$I_\sigma(T) = \int_0^T |f_X(\sigma+it)|^2\,dt, \qquad J_\sigma = \int_{-\infty}^{\infty} |\varphi(\sigma+it)|^2\,dt,$$

where

$$\varphi(s) = \varphi_{X,\tau}(s) = \frac{s-1}{s\cos(s/2\tau)} f_X(s), \qquad \tau > \frac{3}{\pi}. \tag{63}$$

We shall use the estimates for $I_\sigma(T)$ obtained in (61) and (62) to estimate J_σ for $\sigma = 1+\delta$ and $\sigma = \frac{1}{2}$. We shall then apply Lemma 6 to J_σ, in order to obtain an estimate of it which is valid throughout the interval $\frac{1}{2} \leqslant \sigma \leqslant 1+\delta$. From this we shall deduce an estimate for I_σ valid throughout that interval. The passage from I_σ to J_σ and then back again is facilitated by the choice of $\varphi(s)$ made in (63).

In the strip $\frac{1}{2} \leqslant \sigma \leqslant 1+\delta$, $\varphi(s)$ is regular, and satisfies the inequality

$$|\varphi(s)|^2 \leqslant c_1 e^{-\frac{|t|}{\tau}} |f_X(s)|^2, \tag{64}$$

and is therefore bounded for fixed X and τ. Further, for $\frac{1}{2} \leqslant \sigma \leqslant 1+\delta$, $\sigma \neq 1$,

$$J_\sigma \leqslant 2\int_0^\infty c_1 \cdot e^{-t/\tau} |f_X(\sigma+it)|^2\,dt = 2c_1 \int_0^\infty e^{-u} I_\sigma(\tau u)\,du,$$

by partial integration, and the substitution $t = \tau u$. Hence, by (61) and (62), we have

$$J_{1+\delta} < c_2 \int_0^\infty e^{-u}\left(\frac{\tau u}{X} + 1\right)\delta^{-4}\,du < c_3\left(\frac{\tau}{X} + 1\right)\delta^{-4}$$

and

$$J_{\frac{1}{2}} < c_4 \int_0^\infty e^{-u}(\tau u)^{2c}(\tau u + X)\log X\,du$$

$$< c_5\,\tau^{2c}(\tau + X)\log X\,.$$

By Lemma 6, we therefore have

$$J_\sigma < \left(c_3\left(\frac{\tau}{X}+1\right)\delta^{-4}\right)^{\frac{\sigma-\frac{1}{2}}{\frac{1}{2}+\delta}} \cdot \left(c_5\tau^{2c}(\tau+X)\log X\right)^{\frac{1+\delta-\sigma}{\frac{1}{2}+\delta}} \tag{65}$$

for $\frac{1}{2}\leqslant\sigma\leqslant 1+\delta$.

In the strip $\frac{1}{2}\leqslant\sigma\leqslant 1+\delta$, $t\geqslant 1$, $\varphi(s)$ satisfies the inequality

$$|\varphi(s)|^2 \geqslant c_6 e^{-t/\tau}|f_X(s)|^2. \tag{66}$$

Hence (65) implies that for $T>1, \frac{1}{2}\leqslant\sigma\leqslant 1$,

$$c_6 e^{-\frac{T}{\tau}}\int_1^T |f_X(\sigma+it)|^2\,dt$$

$$< X^{\frac{1-2\sigma}{1+2\delta}}\cdot\tau^{\frac{4c(1+\delta-\sigma)}{1+2\delta}}\cdot(\tau+X)\cdot\max(c_3\delta^{-4}, c_5\log X).$$

Taking $\tau=c_7 T$, $\delta=\dfrac{c_8}{\log(T+X)}$, we deduce the lemma, since

$$X^{\frac{1-2\sigma}{1+2\delta}} \leqslant X^{-(1-2\delta)(2\sigma-1)} \leqslant X^{-(2\sigma-1)+2\delta} < e^{2c_9}\ X^{-(2\sigma-1)},$$

and

$$T^{\frac{4c(1+\delta-\sigma)}{1+2\delta}} \leqslant T^{4c(1+\delta-\sigma)} \leqslant T^{4c(1-\sigma)+2\delta} < e^{2c_9}\ T^{4c(1-\sigma)},$$

because $0<c<\frac{1}{2}$.

We are now in a position to prove the following

THEOREM 3 (INGHAM). *If*

$$\zeta(\tfrac{1}{2}+it)=O(t^c), \tag{3}$$

where c is a positive, absolute constant, then

$$N(\sigma,T)=O\left(T^{2(1+2c)(1-\sigma)}\log^5 T\right), \tag{67}$$

uniformly for $\frac{1}{2}\leqslant\sigma\leqslant 1$, *as* $T\to\infty$.

PROOF. As in Lemma 8, we set, for $X>1$,

$$f_X(s)=\zeta(s)\sum_{n<X}\mu(n)n^{-s}-1=\zeta(s)M_X(s)-1. \tag{68}$$

We define $g_X(s)$ and $h_X(s)$ by the relation

$$\begin{aligned} 1-f_X^2(s) &= \zeta(s)M_X(s)(2-M_X(s)\zeta(s)) \\ &= \zeta(s)g_X(s)=h_X(s)=h(s). \end{aligned} \tag{69}$$

Then $g_X(s)$ and $h_X(s)$ are regular except, perhaps, for $s=1$. Further

$$|f_X(s)|^2 \leqslant \left(\sum_{n\geqslant X}\frac{d(n)}{n^2}\right)^2 = O(X^{2\varepsilon-2}) < \frac{1}{2X} < \frac{1}{2},$$

if $\sigma\geqslant 2$, $X>X_0>1$, since $d(n)=O(n^\varepsilon)$, for every $\varepsilon>0$. Therefore $h_X(s)\neq 0$; in fact, $\operatorname{Re} h(s)>\frac{1}{2}$ for $\sigma\geqslant 2$, $X>X_0$.

Let $N_h(\sigma,T)$ denote the number of zeros $\beta+i\gamma$ of $h(s)$ with $\beta\geqslant\sigma$, and $0<\gamma\leqslant T$. If $T_2>T_1$, let

$$N_h(\sigma;T_2,T_1)=N_h(\sigma,T_2)-N_h(\sigma,T_1).$$

Let $X>X_0$, $T>4$; we choose T_1,T_2, so that $3<T_1<4$, $T<T_2<T+1$, and $h(s)$ has no zeros on either of the segments: $\frac{1}{2}\leqslant\sigma\leqslant 2$; $t=T_1$ or $t=T_2$.

Let $\mathscr{R}_u$ be the rectangle bounded by $t=T_1$, $t=T_2$, $\sigma=u(\frac{1}{2}\leqslant u\leqslant 2)$, and $\sigma=2$. We denote the boundary of $\mathscr{R}_u$ by R_u. If there are no zeros of $h(s)$ on R_u, we have

$$\int_{R_u}\frac{h'(s)}{h(s)}\,ds=2\pi i N_h(u;T_2,T_1).$$

If, on the other hand, u is the real part of a zero of $h(s)$ in $\mathscr{R}_{\frac{1}{2}}$, then the integral on the left-hand side is undefined. To obviate this difficulty, we introduce a function $\chi(u)$ of u, for $\frac{1}{2}\leqslant u\leqslant 2$, such that $\chi(u)=0$, if u is the real part of a zero of $h(s)$ in $\mathscr{R}_{\frac{1}{2}}$, and $\chi(u)=1$ otherwise. We then have

$$\int_{R_u}\chi(u)\frac{h'(s)}{h(s)}\,ds=2\pi i\chi(u)N_h(u;T_2,T_1).$$

Let σ_0 be such that $\frac{1}{2}\leqslant\sigma_0\leqslant 1$, and such that there are no zeros of $h(s)$ on R_{σ_0}. Then

$$\begin{aligned}2\pi i\int_{\sigma_0}^{2}N_h(u;T_2,T_1)\,du&=2\pi i\int_{\sigma_0}^{2}\chi(u)N_h(u;T_2,T_1)\,du=\int_{\sigma_0}^{2}du\int_{R_u}\chi(u)\frac{h'(s)}{h(s)}\,ds\\&=\int_{\sigma_0}^{2}du\left\{\left(\int_{u+iT_1}^{2+iT_1}+\int_{2+iT_1}^{2+iT_2}+\int_{2+iT_2}^{u+iT_2}\right)\frac{h'(s)}{h(s)}\,ds\right\}+\\&\quad+\int_{\sigma_0}^{2}du\int_{u+iT_2}^{u+iT_1}\chi(u)\frac{h'(s)}{h(s)}\,ds. \qquad (70)\end{aligned}$$

Now

$$\int_{\sigma_0}^{2} du \int_{u+iT_2}^{u+iT_1} \chi(u) \frac{h'(s)}{h(s)} ds = -i \int_{\sigma_0}^{2} du \int_{T_1}^{T_2} \chi(u) \frac{h'(u+it)}{h(u+it)} dt.$$

The function $\dfrac{h'(u+it)}{h(u+it)}$ in the last integral has only simple poles. Hence the *double* integral

$$\int_{\sigma_0}^{2} \int_{T_1}^{T_2} \chi(u) \frac{h'(u+it)}{h(u+it)} du\, dt$$

is absolutely convergent, and the order of integration can be interchanged. That is

$$\int_{\sigma_0}^{2} du \int_{T_1}^{T_2} \chi(u) \frac{h'(u+it)}{h(u+it)} dt = \int_{T_1}^{T_2} dt \int_{\sigma_0}^{2} \varphi(t) \frac{h'(u+it)}{h(u+it)} \cdot \chi(u)\, du$$

$$= \int_{T_1}^{T_2} dt \int_{\sigma_0}^{2} \varphi(t) \frac{h'(u+it)}{h(u+it)} du,$$

where $\varphi(t)=0$ if t is the imaginary part of a zero of $h(s)$ in $\mathscr{R}_{\frac{1}{2}}$, and $\varphi(t)=1$ otherwise.

Thus (70) leads to the relation

$$2\pi \int_{\sigma_0}^{2} N_h(u; T_2, T_1)\, du = \int_{\sigma_0}^{2} du \left\{ \operatorname{Im} \left(\int_{u+iT_1}^{2+iT_1} + \int_{2+iT_1}^{2+iT_2} + \int_{2+iT_2}^{u+iT_2} \right) \frac{h'(s)}{h(s)} ds \right\} -$$

$$- \int_{T_1}^{T_2} dt \left\{ \operatorname{Re} \int_{\sigma_0}^{2} \varphi(t) \frac{h'(u+it)}{h(u+it)} du \right\}. \qquad (71)$$

The second term on the right-hand side of (71) equals

$$\int_{T_1}^{T_2} \{\log|h(\sigma_0+it)| - \log|h(2+it)|\}\, dt,$$

and we estimate it with the help of Lemma 8, since

$$\log|h(s)| \leqslant \log(1+|f_X(s)|^2) \leqslant |f_X(s)|^2. \qquad (72)$$

In addition, we use the fact that

$$-\log|h(2+it)| \leqslant -\log\{1-|f_X(2+it)|^2\} \leqslant 2|f_X(2+it)|^2 < \frac{1}{X},$$

the last inequality resulting from the estimate $|f_X(2+it)|^2 < \frac{1}{2X} < \frac{1}{2}$ made earlier. Thus we obtain

$$\int_{T_1}^{T_2} \{\log|h(\sigma_0+it)| - \log|h(2+it)|\}\,dt$$

$$< A_1(T+1)^{4c(1-\sigma_0)} \cdot X^{1-2\sigma_0}(T+1+X)\cdot\log^4(T+1+X) + \frac{T}{X}, \quad (73)$$

where A_1 is a constant.

To estimate the first term on the right-hand side of (71), we follow the same method as in the proof of the Riemann-von Mangoldt formula for $\zeta(s)$ in Chapter II.

Let $\mathscr{L}_r$ denote the segment $t=T_r$, $\sigma_0<\sigma<2$, $r=1$ or 2, and m_r the number of points of $\mathscr{L}_r$ at which $\operatorname{Re} h(s)=0$. (We know that $\operatorname{Re} h(s)>0$ for $\sigma=2$, $T_1 \leqslant T \leqslant T_2$). Then, as before,

$$\left|\operatorname{Im} \int_{\mathscr{L}_r} \frac{h'(s)}{h(s)}\,ds\right| \leqslant (m_r+1)\pi, \qquad r=1,2. \quad (74)$$

But m_r is the number of zeros of the function

$$H_r(s) = \tfrac{1}{2}\{h(s+iT_r) + h(s-iT_r)\}$$

on the segment $t=0$, $\sigma_0<\sigma<2$, and therefore cannot exceed the number of zeros of $H_r(s)$ in the circle $|s-2| \leqslant \frac{3}{2}$. Since $H_r(s)$ is regular for $|s-2| \leqslant \frac{7}{4}$, we have, by Lemma A proved in Chapter II,

$$\left(\frac{7}{6}\right)^{m_r} \leqslant \max_{|s-2|\leqslant\frac{7}{4}} \left|\frac{H_r(s)}{H_r(2)}\right| \leqslant \max_{\substack{\sigma\geqslant\frac{1}{4}\\ 1\leqslant t\leqslant T+3}} \frac{|h_X(s)|}{\operatorname{Re} h(2+iT_r)} < (T+X)^{A_2}, \quad (75)$$

where A_2 is an absolute constant, since $\operatorname{Re} h(2+iT_r) > \frac{1}{2}$, because of the definition of $h_X(s)$. Hence the first integral on the right-hand side of (71) is

$$< A_3 \log(T+X). \quad (76)$$

From (76), (73), and (71), we get

$$2\pi \int_{\sigma_0}^{2} N_h(\sigma; T_2, T_1)\,d\sigma < A_4\, T^{4c(1-\sigma_0)}(TX^{1-2\sigma_0} + X^{2(1-\sigma_0)}) \log^4(T+X), \quad (77)$$

since $\frac{T}{X} \leqslant T \cdot X^{1-2\sigma_0}$, and $\log(T+X) \leqslant X^{2(1-\sigma_0)} \log(T+X)$. On the other hand, since $N_h \geqslant N_\zeta$, we have

$$\int_{\sigma_0}^{2} N_h(\sigma; T_2, T_1) d\sigma \geqslant \int_{\sigma_0}^{\sigma_0+\delta} N_\zeta(\sigma; T_2, T_1) d\sigma \geqslant \delta \cdot N_\zeta(\sigma_0+\delta; T_2, T_1),$$

if $0<\delta<1$.

Writing σ for $\sigma_0+\delta$, and noting that $N_\zeta(\sigma, T) < N_\zeta(\sigma; T_2, T_1) + A_5$, we deduce from (77) that

$$N_\zeta(\sigma, T) < A_6 \cdot \frac{1}{\delta} \cdot T^{4c(1-\sigma+\delta)} (T X^{1-2\sigma+2\delta} + X^{2(1-\sigma+\delta)}) \log^4(T+X), \quad (78)$$

for $\frac{1}{2}+\delta \leqslant \sigma \leqslant 1$. Since $T>4$, however, the Riemann-von Mangoldt formula gives

$$N_\zeta(\sigma, T) < A_7 \cdot T \log T \leqslant A_7 \cdot T^{2(1-\sigma+\delta)} \log T, \quad (79)$$

for $\frac{1}{2} \leqslant \sigma \leqslant \frac{1}{2}+\delta$. The theorem now follows from (79) and (78), if we take

$$X = T > \max(X_0, 4), \qquad \delta = \frac{1}{\log T}.$$

§ 6. An application of Chudakov's theorem. It is possible to improve the value of θ in Hoheisel's theorem by making use of Chudakov's theorem that

$$\zeta(s) \neq 0, \quad \text{for} \quad \sigma > 1 - \frac{1}{(\log t)^a}, \qquad t > t_1,$$

where $a<1$, and $t_1 \geqslant 3$. Now condition (34) of Hoheisel's theorem is satisfied with an arbitrarily large A, and an appropriate t_0, while condition (35) is satisfied with $b=2+4c$, $B=5$. Since

$$\lim_{A\to\infty} \left(1 - \frac{1}{b+B/A}\right) = 1 - \frac{1}{b} = \frac{1+4c}{2+4c},$$

we obtain the following

THEOREM 4 (INGHAM). *If*

$$\zeta(\tfrac{1}{2}+it) = O(t^c), \quad (3)$$

for a positive, absolute constant c, then

$$\pi(x+x^\theta) - \pi(x) \sim \frac{x^\theta}{\log x}, \quad (80)$$

and

$$p_{n+1} - p_n = O(p_n^{\theta}), \tag{81}$$

for any fixed θ, *such that*

$$\frac{1+4c}{2+4c} < \theta < 1. \tag{82}$$

REMARKS. (i) It is known, after Hardy and Littlewood, that (3) holds with $c = \frac{1}{6} + \varepsilon$ for every $\varepsilon > 0$, so that we have

$$p_{n+1} - p_n = O(p_n^{\frac{5}{8}+\varepsilon}) \tag{83}$$

for every $\varepsilon > 0$.

(ii) S. H. Min has proved that $c = \frac{15}{92} + \varepsilon$, and W. Haneke that $c = \frac{6}{37} + \varepsilon$, for every $\varepsilon > 0$. These results imply that

$$p_{n+1} - p_n = O\left(p_n^{\frac{38}{61}+\varepsilon}\right), \quad \text{and} \quad p_{n+1} - p_n = O\left(p_n^{\frac{61}{98}+\varepsilon}\right).$$

(iii) If the Lindelöf hypothesis is true, then (3) holds for every $c > 0$, so that $p_{n+1} - p_n = O(p_n^{\frac{1}{2}+\varepsilon})$. On the other hand, it is known that the Riemann hypothesis implies that $p_{n+1} - p_n = O(p_n^{\frac{1}{2}} \log p_n)$.

(iv) The validity of (80) implies that there exists a prime between x and $x + x^{\theta}$, if x is sufficiently large. If in (83), $0 < \varepsilon < \frac{2}{3} - \frac{5}{8} = \frac{1}{24}$, it follows that between n^3 and $(n+1)^3$ there exists a prime, provided that n is sufficiently large. It is an unsolved problem to prove that, for sufficiently large x, there is a prime between x and $x + x^{\frac{1}{2}}$, which would imply that between n^2 and $(n+1)^2$, there is a prime, for n sufficiently large.

Notes on Chapter V

§ 1. Hoheisel's theorem appeared in the *Berliner Sitzungsberichte*, (1930), 580—588; and Ingham's in *Quarterly J. Math.* (Oxford) 8 (1937), 255—266. A. Selberg has shown that (1) is true with a $\theta < \frac{100}{161}$ by a method different from Ingham's. His proof shows that the order of $\zeta(s)$ *on* the line $\sigma = \frac{1}{2}$ is not so relevant to the problem as its order closer to $\sigma = 1$. Selberg's proof is not in print.

Professor E. Bombieri has since informed the author that H. L. Montgomery has proved (1) with $\theta = \frac{3}{5}$ by a new method.

For the Lindelöf hypothesis and its implications, see Titchmarsh's *Zeta-function*, Ch. XIII.

The Hardy-Littlewood value $c = \frac{1}{6} + \varepsilon$ in (3) is obtained, for instance, in Satz 414 of Landau's *Vorlesungen*, II, 61. The value $c = \frac{15}{92} + \varepsilon$ was obtained by S. H. Min, *Trans. Amer. Math. Soc.* 65 (1949), 448—472; and $c = \frac{6}{37} + \varepsilon$ by W. Haneke, *Acta Arithmetica*, 8 (1962/63), 357—430.

§ 2. For Theorem 1 see E. Landau, *Acta Math.* 35 (1912), 271—294. For a slightly different result see his *Vorlesungen*, II, Satz 452. Lemmas 1, 2, 4, 5 are given, for instance, in the same book, 110—115, and Satz 437. Lemma 3 is given in his paper, loc. cit. From the hypothesis $|a_n| < A \log n$ of that lemma, it follows that $(\eta-1)^2 \sum_{n=1}^{\infty} |a_n| n^{-\eta}$ is bounded, so that assumption (9) is non-trivial; take, for example, $f(s)=\zeta'(s)$.

Theorem 1 is related to the so-called *explicit formula* for $\psi(x)$. See Ingham's *Tract*, Theorem 29; also H. Davenport's *Multiplicative number theory*, § 17.

§ 3. This version of Theorem 2 is due to Ingham, see his paper cited above in § 1.

§ 4. Lemma 6 is due to G. H. Hardy, A. E. Ingham, and G. Pólya, *Proc. Royal Soc.* A, 113 (1927), 542—569. Prof. Raghavan Narasimhan says, in a communication, that the following more general result can be proved: "*Let f be holomorphic in the strip $S=\{\sigma_1<\operatorname{Re} s<\sigma_2\}$, $f\not\equiv 0$, and suppose that*

$$J(\sigma) = \int_{-\infty}^{\infty} |f(\sigma+it)|^2\, dt = \lim_{T\to\infty} \int_{-T}^{T} |f(\sigma+it)|^2\, dt$$

converges uniformly for $\sigma_1<\sigma<\sigma_2$. Then $\log J(\sigma)$ is a convex function of σ".

Inequality (50) of Lemma 7 is due to Ingham, *Proc. London Math. Soc.*(2) 27 (1928), 273—300.

Inequality (49) is included in an asymptotic formula stated by S. Ramanujan and proved analytically by B. M. Wilson. An elementary proof is suggested by H. Heilbronn, *Math. Zeitschrift*, 36 (1933), 410, Hilfssatz 20. Another elementary proof results from Lemma B_2 of Ingham's paper, *Proc. London Math. Soc.* (2) 27 (1928), 273—300, if one notes that the result and proof remain valid for $k=0$ if one replaces $\sum_{d|k}$ by $\sum_{d|k,\, d<\sqrt{x}}$.

§§ 5—6. Lemma 8 is Theorem 2 of Ingham's paper, loc. cit. Theorems 4 and 3 are proved by Ingham in the same paper. The proof of Theorem 3 expounded here avoids the use of a theorem of Littlewood, as a result of comments made by Mr. H. Joris and Prof. Raghavan Narasimhan.

For Littlewood's theorem, see J. E. Littlewood, *Proc. Cambridge Phil. Soc.* 22 (1924), 295—318, Theorem 1, and in particular, the foot-note on page 300. He says that several other proofs are possible, without actually giving them. Professor C. L. Siegel comments: "In my opinion the simplest proof goes in the following way. Let $\mathcal{R}$ be a rectangle with orientation parallel to the axes, $\sigma=\sigma_1$ on its left side, $f(s)$ regular on and in $\mathscr{R}$, $f(s)\neq 0$ on the boundary of $\mathscr{R}$. Denote by $\rho=\beta+i\gamma$ the zeros of f inside $\mathscr{R}$. Make a cut from every ρ to the left on $t=\gamma$, and denote the new boundary by $\mathscr{L}$.

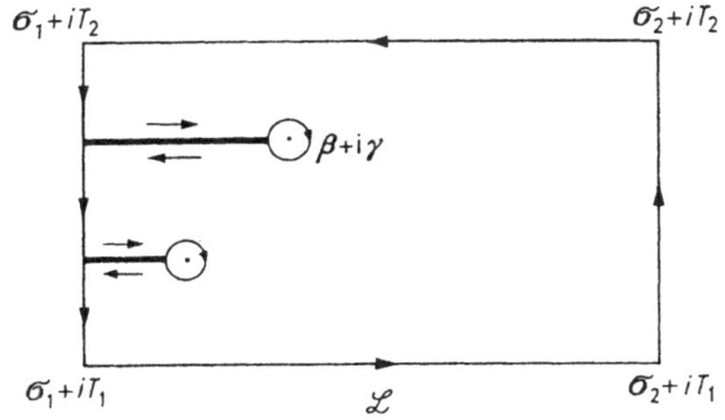

(The figure refers to the case when the imaginary parts of the zeros are all different. Multiple zeros are counted according to their multiplicity). Every branch of $\log f(s)$ is single-valued inside $\mathscr{L}$. By Cauchy's theorem, with any fixed branch of the logarithm,

$$0=\int_{\mathscr{L}}\log f(s)\,ds=\int_{\mathscr{R}}\log f(s)\,ds+2\pi i\sum_{\rho}(\beta-\sigma_1)\,.$$

On the other hand, if $N(\sigma_0)$ denotes the number of zeros of $f(s)$ in the rectangle $\sigma_0\leqslant\sigma\leqslant\sigma_2$, $T_1\leqslant t\leqslant T_2$, where $\sigma_1\leqslant\sigma_0\leqslant\sigma_2$, then

$$\sum_{\rho}(\beta-\sigma_1)=\int_{\sigma_1}^{\sigma_2}N(\sigma)\,d\sigma\,,$$

which gives the required result. It does not matter that $\log f(s)$ jumps by a multiple of $2\pi i$ at the points $\sigma_1+i\gamma$ on the left side of $\mathscr{R}$. The formula remains valid, if $f(s)$ has zeros on this left side.

Probably this is one of Littlewood's 'other proofs', but I know of no others.

Perhaps the proof is still simpler, if one considers first the case $f(s)=s-\rho$, then the case of a function without zeros, and finally uses the additivity of the logarithm."

That the Riemann hypothesis implies the result $p_{n+1}-p_n=O(p_n^{\frac{1}{2}}\log p_n)$ was proved by H. Cramér, *Arkiv. for Mat., Astr., och Fys.* 15 (1921), No. 5; *Acta Arithmetica*, 2 (1936), 23—46.

Theorem 3 is not very significant over the range $\frac{1}{2} \leqslant \sigma \leqslant \frac{1+4c}{2+4c}$ where the exponent of T is greater than 1. Ingham, in the paper already cited, gave a companion theorem to the effect that

$$N(\sigma, T) = O(T^{(1+2\sigma)(1-\sigma)} \log^5 T)$$

uniformly for $\frac{1}{2} \leqslant \sigma \leqslant 1$, as $T \to \infty$. This is better than Theorem 3 for $\frac{1}{2} < \sigma < \frac{1}{2} + 2c$, but worse for $\frac{1}{2} + 2c < \sigma < 1$. In a later paper, *Quarterly J. Math.* (Oxford) 11 (1940), 291—292, he proved that

$$N(\sigma, T) = O(T^{3(1-\sigma)/(2-\sigma)} \log^5 T),$$

by using a convexity theorem, in two variables, due to R. M. Gabriel, *J. London Math. Soc.* 2 (1927), 112—117. A. Selberg has proved a better result to the effect that

$$N(\sigma, T) = O(T^{1-\frac{1}{4}(\sigma-\frac{1}{2})} \log T),$$

uniformly for $\frac{1}{2} \leqslant \sigma \leqslant 1$, in *Arch. for Math. og Naturv.* B, 48 (1946), No. 5. See Titchmarsh's *Zeta-function*, 200—204.

For comments on the Hoheisel-Ingham theorems, see Davenport's *Multiplicative number theory*, 188.

Chapter VI

Dirichlet's L-functions and Siegel's theorem

§ 1. Characters and L-functions. A character of a finite abelian group G is a complex-valued function, not identically zero, defined on the group, such that if $A \in G$, $B \in G$, then $\chi(AB) = \chi(A) \cdot \chi(B)$, where AB is the group-composite of A and B. If E denotes the unit element of G, and A^{-1} the group inverse of $A \in G$, we assume as known the following properties of characters:

(1) if G is of order h, $\chi(A)$ is an hth root of unity, and the character defined by the property $\chi_1(A) = 1$ for *every* $A \in G$, is called the *principal character* of G;

(2) an abelian group of order h has exactly h characters;

(3) $\chi(E) = 1$, for every character χ of G, and given any $A \in G$, $A \neq E$, there exists a character χ, such that $\chi(A) \neq 1$;

(4) the characters of G again form a finite, multiplicative, abelian group of which the principal character is the unit element;

(5) $$\sum_A \chi(A) = \begin{cases} h, & \text{if } \chi = \chi_1, \\ 0, & \text{if } \chi \neq \chi_1; \end{cases} \qquad \sum_\chi \chi(A) = \begin{cases} h, & \text{if } A = E, \\ 0, & \text{if } A \neq E. \end{cases}$$

Given a *positive* integer k, the prime residue classes modulo k form a multiplicative, abelian group of order $h = \varphi(k)$, where φ denotes Euler's arithmetical function. We can take G to be this group, and consider its characters. We can carry over the definition of character from the prime residue classes mod k to the integers belonging to those classes, by the requirement that if A is such a class, and $a \in A$, then $\chi(a) = \chi(A)$. It follows that if $a \equiv b \pmod k$, then $\chi(a) = \chi(b)$; and if $(a,k) = (b,k) = 1$, then $\chi(ab) = \chi(a)\chi(b)$. Since $\chi(A) \neq 0$, if A is any prime residue class mod k, it follows that $\chi(a) \neq 0$, if $(a,k) = 1$. We can carry over the definition of character from the set of integers prime to k to the set of *all* integers by the requirement that $\chi(a) = 0$ if $(a,k) > 1$.

A *character modulo* k is therefore a complex-valued function, of absolute value less than or equal to 1, defined on the set of all integers, such that $\chi(a)=\chi(b)$ if $a\equiv b(\bmod k)$; $\chi(ab)=\chi(a)\chi(b)$; $\chi(a)=0$, if $(a,k)>1$; and $\chi(a)\neq 0$, if $(a,k)=1$. There exist $\varphi(k)$ characters modulo k, which form a multiplicative abelian group isomorphic to the group of prime residue classes modulo k. The unit element of this group is the principal character χ_1, which is such that $\chi_1(a)=1$ if $(a,k)=1$. Further we have the *relations of orthogonality*, namely

$$\sum_{n(\bmod k)} \chi(n) = \begin{cases} \varphi(k), & \text{if } \chi=\chi_1, \\ 0, & \text{if } \chi\neq\chi_1. \end{cases} \qquad \sum_{\chi} \chi(n) = \begin{cases} \varphi(k), & \text{if } n\equiv 1(\bmod k), \\ 0, & \text{if } n\not\equiv 1(\bmod k). \end{cases}$$

With each of the $\varphi(k)$ characters modulo k, we associate a *Dirichlet series* $\sum_{n=1}^{\infty} \chi(n)n^{-s}$, $s=\sigma+it$. Since $|\chi(n)|\leqslant 1$, this series converges absolutely for $\sigma>1$, and we denote its sum-function by $L(s,\chi)$. For different characters χ we get different functions $L(s,\chi)$, called *Dirichlet's L-functions.*

Since each character χ is a completely multiplicative arithmetical function, we have, by *Euler's identity*,

$$L(s,\chi) = \sum_{n=1}^{\infty} \frac{\chi(n)}{n^s} = \prod_p \left(1 - \frac{\chi(p)}{p^s}\right)^{-1}, \qquad \sigma>1,$$

where p runs through all the primes. The product is absolutely convergent for $\sigma>1$, and has non-zero factors. Hence $L(s,\chi)\neq 0$ for $\sigma>1$.

If χ_1 is the principal character modulo k, $\chi_1(a)=1$ if $(a,k)=1$, and $\chi_1(a)=0$ if $(a,k)>1$. Hence

$$L(s,\chi_1) = \prod_p (1-p^{-s})^{-1} \prod_{p|k} (1-p^{-s}), \qquad \sigma>1.$$

Since

$$\zeta(s) = \prod_p (1-p^{-s})^{-1}, \qquad \sigma>1,$$

where ζ denotes Riemann's zeta-function, we have

$$L(s,\chi_1) = \zeta(s) \prod_{p|k} (1-p^{-s}), \qquad \sigma>1.$$

Using the properties of ζ, we infer that $L(s,\chi_1)$ is a regular function of s all over the s-plane except for a simple pole at $s=1$, the residue at the pole being $\dfrac{\varphi(k)}{k}$, since $\varphi(k)=k\prod_{p|k}(1-p^{-1})$. If $\chi\neq\chi_1$, $L(s,\chi)$ is an entire function of s. For $\sigma>1$,

$$L(s,\chi) = \sum_{a=1}^{\infty} \frac{\chi(a)}{a^s} = \sum_{a=1}^{k-1} \chi(a) \sum_{n=0}^{\infty} \frac{1}{(kn+a)^s} = \frac{1}{k^s} \sum_{a=1}^{k-1} \chi(a) \sum_{n=0}^{\infty} \frac{1}{\left(n+\dfrac{a}{k}\right)^s}.$$

The series $\sum_{n=1}^{\infty}(n+\delta)^{-s}$, $0\leqslant\delta<1$, defines a regular function of s all over the s-plane, except for a simple pole at $s=1$ with residue 1. The proof is similar to the case of $\zeta(s)$. (See Chapter II, § 1). Since $\sum_{a=1}^{k-1}\chi(a)=0$, it follows that $L(s,\chi)$ is entire. This can also be deduced from the *functional equation* for $L(s,\chi)$ which we later establish.

It is known that $L(1,\chi)\neq 0$, for every character χ modulo k, and imitating the argument used for $\zeta(s)$, one can prove that $L(s,\chi)\neq 0$ for $\sigma=1$.

§ 2. Zeros of L-functions. Let k and l be integers, $k>0$, $(k,l)=1$, and let

$$\pi(x;k,l)=\sum_{\substack{p\leqslant x\\ p\equiv l\,(\mathrm{mod}\,k)}}1,$$

where p denotes a prime. We have $\pi(x;1,1)=\pi(x)$, in the notation of Chapter I. Dirichlet's theorem on the existence of an infinity of primes in any arithmetical progression implies that, given k and l, $\pi(x;k,l)\to\infty$ as $x\to\infty$. The proof follows from the property that $L(1,\chi)\neq 0$ for every character χ. The prime number theorem for arithmetical progressions takes the form

$$\pi(x;k,l)\sim\frac{\operatorname{li}x}{\varphi(k)},\quad \text{as}\quad x\to\infty,$$

where φ denotes Euler's function. As already remarked (see the Notes on Chapter I), this can be proved by Selberg's method. For the error term in this asymptotic relation, Landau obtained the estimate

$$\pi(x;k,l)-\frac{\operatorname{li}x}{\varphi(k)}=O\left(x\,e^{-c\sqrt{\log x\log\log x}}\right),\qquad c>0,$$

where c is independent of k and l. The proof is based on an extension, to L-functions, of Littlewood's theorem on the zeros of $\zeta(s)$, namely

$$L(s,\chi)\neq 0,\quad \text{for}\quad t\geqslant\tau(k),\qquad \sigma>1-\frac{A\log\log t}{\log t},$$

where A is a positive, absolute constant, and $\tau(k)>3$.

Chudakov's theorem on the zero-free region of $\zeta(s)$ also carries over to L-functions:

$$L(s,\chi)\neq 0,\quad \text{for}\quad t\geqslant\tau(k),\qquad \sigma>1-\frac{1}{(\log t)^{a}},$$

for a certain number $a<1$, and $\tau(k)>3$. This leads to a further improvement in the estimation of the error term.

What we here seek is an estimate of $\pi(x;k,l)-\frac{\operatorname{li} x}{\varphi(k)}$, which is uniform both in k and in l. An estimate which is uniform in l alone is trivial. For the function $\pi(x;k,l)$, for any given k, depends only on x and on the residue class of $l \bmod k$. So there are only $\varphi(k)$ such functions. An estimate which holds *uniformly* in k and l, provided that $1 \leqslant k \leqslant \log^\alpha x$, $x>1$, where α is any positive number, however large, results from a knowledge of the location of the *real* zeros of $L(s,\chi)$, where χ is a *real*, *non-principal* character modulo k. The first advance in this direction, as well as the best up to date, is due to C. L. Siegel. It follows from Siegel's work that given any $\varepsilon>0$, there exists a number $A=A(\varepsilon)>0$, such that if $k>A$, and χ a real, non-principal character modulo k, then $L(s,\chi)$ has no zero on the real axis between $1-k^{-\varepsilon}$ and 1. We shall prove this result. It implies, as A. Walfisz has shown, that

$$\pi(x;k,l)-\frac{\operatorname{li} x}{\varphi(k)}=O\left(x\,e^{-c\sqrt{\log x}}\right),$$

as $x\to\infty$, *uniformly* for $3\leqslant k\leqslant \log^\alpha x$, where α is any positive number however large, and c a positive, absolute constant.

§ 3. Proper characters. We consider the class of all characters modulo k, and distinguish the *proper* characters from the rest, called the *improper* characters. The study of L-functions with improper characters can be reduced to that of L-functions with proper characters.

Let χ be a character modulo k, and $k|l$. Then χ is *extended to* a character ψ modulo l by the definition: $\psi(n)=0$ if $(n,l)>1$, and $\psi(n)=\chi(n)$, if $(n,l)=1$. We then say that ψ is *derived from* the character χ modulo k. If χ modulo k is extended to χ' modulo k', and χ' is extended to ψ modulo l, χ is extended to ψ. The following elementary result is useful.

If k is a divisor of l, then every integer a which is prime to k is congruent modulo k to an integer x prime to l. For let l_0 be the product of those prime factors of l which do not divide k. Then $(l_0,k)=1$, so that integers y',z' exist such that $y'k+z'l_0=1$. Hence $yk+zl_0=a-1$ for some integers y,z. If we set $x=a-yk=1+zl_0$, then $(x,k)=1$, since $(a,k)=1$; and $(x,l_0)=1$. Hence $(x,l)=1$, and $x\equiv a \pmod{k}$.

It follows that *a character ψ modulo l is derived from at most one character modulo k*. For if χ' modulo k is another character from which ψ is derived, and $(n,k)=1$, then $\chi'(n)=\chi(n)$. This follows from the fact that there exists an integer m, such that $m \equiv n \pmod k$, and $(m,l)=1$. Therefore $\chi'(n)=\chi'(m)=\psi(m)=\chi(m)=\chi(n)$.

Further, *a character ψ modulo l is derived from a character modulo k, if and only if $\psi(b)=\psi(c)$ for $(bc,l)=1$, $b \equiv c \pmod k$*. It is clear that *if* ψ modulo l is derived from χ modulo k, then $\psi(b)=\psi(c)$ for $(bc,l)=1$, $b \equiv c \pmod k$. Conversely, if $\psi(b)=\psi(c)$ for $(bc,l)=1$, $b \equiv c \pmod k$, then we can define a character χ modulo k as follows: if $k|l$ and $(a,k)=1$, then there exists a b such that $b \equiv a \pmod k$, and $(b,l)=1$, and we define $\chi(a)=\psi(b)$. This definition does not depend on the choice of b, for if $(c,l)=1$, and $c \equiv a \pmod k$, then $c \equiv b \pmod k$ and $\psi(b)=\psi(c)$ by assumption. Clearly ψ modulo l is derived from χ modulo k so defined.

A character modulo l is *proper*, if it is not derived from any character modulo k for $k<l$; otherwise it is *improper*.

If a character ψ modulo l is derived both from a character modulo k, and from a character modulo m, then it is derived from a character χ modulo (k,m). For let $(ab,l)=1$, $a \equiv b \pmod{(k,m)}$. Then there exist integers x,y such that $a+kx=b+my=z$, say, whence $(z,k)=1$, $(z,m)=1$, $a \equiv z \pmod k$, $b \equiv z \pmod m$. There exists z' such that $(z',l)=1$, $z' \equiv z$ $(\bmod \{k,m\})$, where $\{k,m\}$ is the least common multiple of k and m. Hence $\psi(a)=\psi(z')=\psi(b)$.

There exists a least integer $f>0$, such that ψ modulo l is derived from a character modulo f. The integer f is called the *conductor* of the character ψ modulo l. It follows that f is the smallest integer, greater than zero, such that $\psi(a)=1$, for every integer a with the properties $(a,l)=1$, $a \equiv 1 \pmod f$.

Every character is derived from a unique proper character. For if the character ψ modulo l has conductor f, it is derived from a character χ modulo f, say, which *is* proper, for otherwise χ modulo f is derived from another character χ' modulo g, $g<f$, and that would imply that ψ modulo l is itself derived from χ' modulo g, $g<f$. This contradicts the fact that f is the conductor of ψ modulo l. To show the uniqueness of χ modulo f, let χ' modulo f' be any other proper character from which ψ modulo l is derived. Then ψ modulo l is derived from a character χ'' modulo (f,f'). It follows that χ' modulo f' is derived from χ'' modulo (f,f'), and since χ' is proper, $(f,f')=f'$. Similarly $(f,f')=f$. Hence $f=f'$, and therefore $\chi=\chi'$.

Two characters are said to be *equivalent*, if they are derived from the same proper character; otherwise *non-equivalent*. Thus if χ modulo k and χ' modulo k' are equivalent, then $\chi(n)=\chi'(n)$ for $(n,kk')=1$.

Given two characters χ modulo k, and χ' modulo k', we define the *product* $\chi\chi'$ by the requirement: $\chi\chi'(n)=\chi(n)\chi'(n)$ for every integer n. Clearly $\chi\chi'$ modulo kk' is a character. Hence *the product of two characters is a character.*

With every character χ modulo k, with conductor f, we can associate a unique proper character χ' modulo f, such that $\chi=\chi_1\cdot\chi'$, where χ_1 is the principal character modulo k. Clearly χ and χ' are equivalent.

We shall see later that the functions $L(s,\chi)$ corresponding to equivalent, non-principal characters have the same zeros in the half-plane $\sigma>0$.

The principal character modulo k has conductor 1, so that it is proper if $k=1$, and improper if $k>1$. There is no character modulo k which has conductor 2, since there is no proper character modulo 2, the only character modulo 2 being the principal character. If k is an odd prime, then every non-principal character is proper.

If χ is an *improper* character modulo k, and

$$L(s,\chi)=\prod_p (1-\chi(p)p^{-s})^{-1}, \qquad \sigma>1,$$

then we can associate with χ a proper character χ' modulo f, such that $\chi=\chi_1\cdot\chi'$, χ_1 being the principal character modulo k. Hence

$$\begin{aligned} L(s,\chi) &= \prod_p (1-\chi'(p)p^{-s})^{-1}\prod_{p\mid k}(1-\chi'(p)p^{-s}) \\ &= \prod_{p\mid k}(1-\chi'(p)p^{-s})\cdot\sum_{n=1}^{\infty}\frac{\chi'(n)}{n^s} \\ &= \prod_{\nu=1}^{r}\left(1-\frac{\varepsilon_\nu}{p_\nu^s}\right)L(s,\chi'), \qquad (1)\end{aligned}$$

where $L(s,\chi')$ is an L-function with a proper character χ' modulo f, r is the number of primes which divide k but not f (since $\chi'(p)=0$ if $p\mid f$), and ε_ν is a root of unity. If $f=1$, we get Riemann's zeta-function $\zeta(s)$ instead of $L(s,\chi')$. *We may, in the sequel, suppose that $k>1$, and χ proper, hence non-principal, thus $k\geqslant 3$.*

Let

$$G(\chi,\eta)=\sum_{l\,(\mathrm{mod}\,k)}\chi(l)\eta^l, \qquad (2)$$

where χ is a character modulo k, η a kth root of unity, and the summation is over l running through a complete residue system mod k. We shall show that *if χ is proper* (as already supposed), *then*

$$G(\chi,\eta^r)=\bar{\chi}(r)G(\chi,\eta), \qquad (3)$$

for every integer r, where $\bar{\chi}(r)$ denotes the complex conjugate of $\chi(r)$.

CASE (i). Let $(r,k)=1$. Then $\chi(rl)=\chi(r)\chi(l)=\chi(l)\cdot\dfrac{1}{\overline{\chi}(r)}$, so that $\chi(l)=\chi(rl)\overline{\chi}(r)$, which gives

$$G(\chi,\eta^r)=\overline{\chi}(r)\sum_{l\,(\mathrm{mod}\,k)}\chi(rl)\eta^{rl}=\overline{\chi}(r)\,G(\chi,\eta),\tag{4}$$

since $r\,l$ runs through a complete residue system mod k as l does.

CASE (ii). Let η be an *imprimitive* k^{th} root of unity, so that $\eta^d=1$ for a proper divisor d of k. Since χ modulo k is proper, we can find an integer m, such that $(m,k)=1$, $m\equiv 1(\mathrm{mod}\,d)$, and $\chi(m)\neq 1$. Thus $\eta^m=\eta$, and (4) implies that $G(\chi,\eta)=0$.

Now let $(r,k)>1$. Then η^r is an imprimitive k^{th} root of unity, so that, if χ is proper, then $G(\chi,\eta^r)=0$, and since $\chi(r)=0$, (3) follows again.

We shall next show that *if χ is proper, and η is primitive, then*

$$|G(\chi,\eta)|=\sqrt{k}.\tag{5}$$

We have

$$\sum_{n\,(\mathrm{mod}\,k)}|G(\chi,\eta^n)|^2=\varphi(k)\,|G(\chi,\eta)|^2,$$

since, by (3), the non-zero terms of the sum are all equal, and $\varphi(k)$ in number. But

$$\begin{aligned}\sum_{n\,(\mathrm{mod}\,k)}|G(\chi,\eta^n)|^2&=\sum_{n\,(\mathrm{mod}\,k)}\ \sum_{l,r}\chi(l)\overline{\chi}(r)\eta^{n(l-r)}\\&=\sum_{l,r}\chi(l)\overline{\chi}(r)\sum_{n\,(\mathrm{mod}\,k)}\eta^{n(l-r)}\\&=k\sum_{l\equiv r\,(\mathrm{mod}\,k)}\chi(l)\overline{\chi}(r)=k\,\varphi(k),\end{aligned}$$

from which (5) follows.

§ 4. The functional equation of $L(s,\chi)$.

We obtained, in Chapter II, the following theta-relation

$$\sum_{n=-\infty}^{\infty}e^{-\pi n^2 t}=\frac{1}{\sqrt{t}}\sum_{n=-\infty}^{\infty}e^{-\pi n^2/t},\qquad t>0,\tag{6}$$

as a special case of Poisson's summation formula. Similarly one can obtain, for any real α,

$$\sum_{m=-\infty}^{\infty}e^{-\pi(m+\alpha)^2\tau}=\frac{1}{\sqrt{\tau}}\sum_{m=-\infty}^{\infty}e^{-\pi m^2/\tau+2\pi i m\alpha},\qquad \tau>0.\tag{7}$$

If χ is a character modulo k, then $(\chi(-1))^2 = \chi(1) = 1$, so that $\chi(-1) = \pm 1$. We consider a generalization of (6) in these two cases separately.

THEOREM 1. *Let χ be a proper character modulo k, $k > 1$. Let*

$$\eta = e^{2\pi i/k}, \qquad G(\chi, \eta) = \sum_{r=0}^{k-1} \chi(r) e^{\frac{2\pi i r}{k}}.$$

If $\chi(-1) = 1$, then

$$\psi(\tau, \chi) = \frac{\varepsilon(\chi)}{\sqrt{\tau}} \psi\left(\frac{1}{\tau}, \bar{\chi}\right), \tag{8}$$

where

$$\psi(\tau, \chi) = 2 \sum_{n=1}^{\infty} \chi(n) e^{-\pi n^2 \tau/k}, \qquad \tau > 0, \tag{8$'$}$$

and

$$\varepsilon(\chi) = \frac{G(\chi, \eta)}{\sqrt{k}}, \qquad \sqrt{k} > 0. \tag{8$''$}$$

If, on the other hand, $\chi(-1) = -1$, then

$$\psi_1(\tau, \chi) = \frac{\varepsilon_1(\chi)}{\tau\sqrt{\tau}} \psi_1\left(\frac{1}{\tau}, \bar{\chi}\right), \tag{9}$$

where

$$\psi_1(\tau, \chi) = 2 \sum_{n=1}^{\infty} n\chi(n) e^{-\pi n^2 \tau/k}, \qquad \tau > 0, \tag{9$'$}$$

and

$$\varepsilon_1(x) = -i \frac{G(\chi, \eta)}{\sqrt{k}}, \qquad \sqrt{k} > 0. \tag{9$''$}$$

PROOF. We first consider the case $\chi(-1) = 1$. Since $\chi(n) = \chi(-n)\chi(-1) = \chi(-n)$, and $\chi(0) = 0$, we have

$$\psi(\tau, \chi) = \sum_{n=-\infty}^{\infty} \chi(n) e^{-\pi n^2 \tau/k}. \tag{10}$$

If we set $n = mk + r$, $0 \leqslant r \leqslant k-1$, then

$$\psi(\tau, \chi) = \sum_{r=0}^{k-1} \chi(r) \sum_{m=-\infty}^{\infty} e^{-\pi\left(m + \frac{r}{k}\right)^2 k\tau}, \tag{11}$$

and if we use (7), we get

$$\psi(\tau,\chi) = \frac{1}{\sqrt{k\tau}} \sum_{m=-\infty}^{\infty} e^{-\pi m^2/(k\tau)} \sum_{r=0}^{k-1} \chi(r) e^{\frac{2\pi i m r}{k}}$$

$$= \frac{1}{\sqrt{k\tau}} \sum_{m=-\infty}^{\infty} e^{-\pi m^2/(k\tau)} G(\chi,\eta^m),$$

and, because of (3), this leads to

$$\psi(\tau,\chi) = \frac{1}{\sqrt{k\tau}} G(\chi,\eta) \sum_{m=-\infty}^{\infty} \overline{\chi}(m) e^{-\pi m^2/(k\tau)}$$

$$= \frac{1}{\sqrt{k\tau}} G(\chi,\eta) \psi\left(\frac{1}{\tau}, \overline{\chi}\right)$$

$$= \frac{\varepsilon(\chi)}{\sqrt{\tau}} \psi\left(\frac{1}{\tau}, \overline{\chi}\right). \tag{12}$$

By repeating the transformation $\tau \to \frac{1}{\tau}$, we get

$$\psi(\tau,\chi) = \frac{\varepsilon(\chi)}{\sqrt{\tau}} \cdot \varepsilon(\overline{\chi}) \cdot \sqrt{\tau} \cdot \psi(\tau,\chi),$$

or $\varepsilon(\chi)\cdot\varepsilon(\overline{\chi}) = 1$. But $\varepsilon(\chi)$ is the complex conjugate of $\varepsilon(\overline{\chi})$, since

$$\varepsilon(\chi) = \frac{1}{\sqrt{k}} \sum_{n=1}^{k} \chi(n) e^{\frac{2\pi i n}{k}} = \frac{1}{\sqrt{k}} \sum_{n=1}^{k} \chi(-n) e^{-\frac{2\pi i n}{k}} = \frac{1}{\sqrt{k}} \sum_{n=1}^{k} \chi(n) e^{-\frac{2\pi i n}{k}}.$$

Hence $|\varepsilon(\chi)| = 1$. This was established directly in (5).

We next consider the case $\chi(-1) = -1$. Then

$$\psi_1(\tau,\chi) = \sum_{n=-\infty}^{\infty} n\chi(n) e^{-\pi n^2 \tau/k}, \tag{13}$$

and writing $n = mk + r$, $0 \leqslant r \leqslant k-1$, we get

$$\psi_1(\tau,\chi) = \sum_{r=0}^{k-1} \chi(r) \sum_{m=-\infty}^{\infty} (mk+r) e^{-\pi\left(m+\frac{r}{k}\right)^2 k\tau}. \tag{13$'$}$$

Relation (7) can be differentiated with respect to α, since for a fixed $\tau > 0$, the differentiated series, on either side, converge uniformly in every finite α-interval. Hence

$$-\sum_{m=-\infty}^{\infty} (m+\alpha) e^{-(m+\alpha)^2 \pi\tau} = \frac{i}{\tau\sqrt{\tau}} \sum_{m=-\infty}^{\infty} m e^{-\pi m^2/\tau + 2\pi i m \alpha}.$$

Using this in (13)′, we obtain

$$\psi_1(\tau,\chi) = -\frac{i}{\tau\sqrt{\tau k}} \sum_{m=-\infty}^{\infty} m e^{-\pi m^2/(k\tau)} \sum_{r=0}^{k-1} \chi(r) e^{2\pi i m r/k}$$

$$= \frac{1}{i\tau\sqrt{\tau k}} \sum_{m=-\infty}^{\infty} m e^{-m^2\pi/(k\tau)} G(\chi,\eta^m)$$

$$= \frac{1}{i\tau\sqrt{\tau k}} \sum_{m=-\infty}^{\infty} \overline{\chi}(m) m e^{-m^2\pi/(k\tau)} G(\chi,\eta),$$

by (3); that is

$$\psi_1(\tau,\chi) = \frac{\varepsilon_1(\chi)}{\tau\sqrt{\tau}} \psi_1\left(\frac{1}{\tau}, \overline{\chi}\right). \tag{14}$$

By repeating the transformation $\tau \to \frac{1}{\tau}$, as before, we obtain $\varepsilon_1(\chi)\cdot\varepsilon_1(\overline{\chi}) = 1$. However, $\varepsilon_1(\chi)$ is the conjugate of $\varepsilon_1(\overline{\chi})$, for $\chi(-1) = -1$, and

$$\varepsilon_1(\chi) = -\frac{i}{\sqrt{k}} \sum_{r=0}^{k-1} \chi(r) e^{\frac{2\pi i r}{k}}.$$

Hence $|\varepsilon_1(\chi)| = 1$, which was directly proved in (5). Now (12) and (14) give the theorem.

We shall use Theorem 1 to establish the *functional equation* of $L(s,\chi)$.

THEOREM 2. *Let χ be a proper, non-principal, character modulo k. Let*

$$a = \begin{cases} 0, & \text{if } \chi(-1) = 1, \\ 1, & \text{if } \chi(-1) = -1, \end{cases} \tag{15}$$

and

$$E(\chi) = \begin{cases} \varepsilon(\chi), & \text{if } a = 0, \\ \varepsilon_1(\chi), & \text{if } a = 1, \end{cases} \tag{16}$$

where

$$\varepsilon(\chi) = \frac{G(\chi,\eta)}{\sqrt{k}}, \quad \varepsilon_1(\chi) = -i\frac{G(\chi,\eta)}{\sqrt{k}}, \quad \sqrt{k} > 0,$$

$$G(\chi,\eta) = \sum_{l \,(\mathrm{mod}\, k)} \chi(l)\eta^l, \quad \eta = e^{2\pi i/k}.$$

Then

$$\xi(s,\chi)=E(\chi)\,\xi(1-s,\bar{\chi}), \tag{17}$$

where

$$\xi(s,\chi)=\left(\frac{\pi}{k}\right)^{-\frac{1}{2}(s+a)}\Gamma(\tfrac{1}{2}s+\tfrac{1}{2}a)\,L(s,\chi). \tag{18}$$

PROOF. We first consider the case $\chi(-1)=1$. The gamma-function formula

$$\left(\frac{\pi}{k}\right)^{-\frac{1}{2}s}\Gamma\left(\frac{1}{2}s\right)n^{-s}=\int_0^\infty e^{-\frac{\pi n^2 u}{k}}u^{\frac{1}{2}s-1}\,du, \qquad \sigma>1,$$

leads to

$$\xi(s,\chi)=\int_0^\infty\sum_{n=1}^\infty \chi(n)e^{-\frac{\pi n^2 u}{k}}u^{\frac{1}{2}s-1}\,du, \qquad \sigma>1,$$

$$=\frac{1}{2}\int_0^\infty \psi(u,\chi)u^{\frac{1}{2}s-1}\,du \qquad \text{(by (8)')}$$

$$=\frac{1}{2}\left(\int_0^1+\int_1^\infty\right)\psi(u,\chi)u^{\frac{1}{2}s-1}\,du.$$

In the first integral on the right-hand side, we use (12), and obtain

$$\xi(s,\chi)=\frac{\varepsilon(\chi)}{2}\int_0^1\frac{\psi\left(\frac{1}{u},\bar{\chi}\right)}{u^{\frac{1}{2}}}\cdot u^{\frac{1}{2}s-1}\,du+\frac{1}{2}\int_1^\infty\psi(u,\chi)u^{\frac{1}{2}s}\frac{du}{u}$$

$$=\frac{\varepsilon(\chi)}{2}\int_1^\infty\psi(u,\bar{\chi})\cdot u^{\frac{1}{2}}\,u^{1-\frac{1}{2}s}\frac{du}{u^2}+\frac{1}{2}\int_1^\infty\psi(u,\chi)u^{\frac{1}{2}s}\cdot\frac{du}{u}.$$

The integrals on the right-hand side converge uniformly (see Theorem 2, Chapter II), and represent entire functions of s, and

$$\varepsilon(\chi)\,\xi(1-s,\bar{\chi})=\xi(s,\chi),$$

as claimed.

If $\chi(-1)=-1$, we start from the formulas

$$\left(\frac{\pi}{k}\right)^{-\frac{1}{2}(s+1)} \Gamma(\tfrac{1}{2}+\tfrac{1}{2}s)n^{-s} = \int_0^\infty n\, e^{-\pi n^2 u/k} u^{\frac{1}{2}(s+1)-1}\,du, \qquad \sigma>1,$$

and

$$\xi(s,\chi) = \left(\frac{\pi}{k}\right)^{-\frac{1}{2}(s+1)} \Gamma(\tfrac{1}{2}+\tfrac{1}{2}s)\,L(s,\chi).$$

By (9)′ we have

$$\xi(s,\chi) = \frac{1}{2}\int_0^\infty \psi_1(u,\chi)u^{\frac{1}{2}(s+1)}\cdot\frac{du}{u} = \int_0^1 + \int_1^\infty.$$

If we use (14) in the first integral, and make a change of variable $u \to \dfrac{1}{u}$, we see, as before, that $\xi(s,\chi)$ is an entire function of s, and

$$\xi(s,\chi) = \varepsilon_1(\chi)\,\xi(1-s,\overline{\chi}),$$

thus completing the proof of the theorem.

Remarks.

(i) We have seen in Chapter II that the function

$$\xi(s) = \tfrac{1}{2}s(s-1)\pi^{-\frac{1}{2}s}\Gamma(\tfrac{1}{2}s)\zeta(s)$$

is entire, and satisfies the equation $\xi(s)=\xi(1-s)$. If $k=1$, $\chi\equiv1$, then $\xi(s,\chi)=\xi(s)$.

(ii) Since $L(s,\chi)\neq0$ for $\sigma>1$, $\xi(s,\chi)\neq0$, and $\xi(s,\overline{\chi})\neq0$, for $\sigma>1$. By the functional equation (17), $\xi(s,\chi)\neq0$ for $\sigma<0$, so that all the zeros of $\xi(s,\chi)$ must lie in the strip $0\leqslant\sigma\leqslant1$.

It is known that $L(1,\chi)\neq0$, so that $\xi(1,\chi)\neq0$ and $\xi(1,\overline{\chi})\neq0$, hence $\xi(0,\chi)\neq0$.

(iii) If χ is a proper, non-principal character modulo k, and $\chi(-1)=1$, then $L(s,\chi)$ has simple zeros at $s=0, -2, -4, \ldots$. If $\chi(-1)=-1$, then $L(s,\chi)$ has simple zeros at $s=-1, -3, -5, \ldots$. These are called the *trivial zeros*.

(iv) Taking Remark (i) into account, as well as (1), we can express $L(s,\chi)$, for *any* character χ modulo k, as follows:

$$L(s,\chi) = \frac{2^b}{s^b(s-1)^b}\cdot\prod_{\nu=1}^{l}\left(1-\frac{\varepsilon_\nu}{p_\nu^s}\right)\left(\frac{\pi}{f}\right)^{\frac{1}{2}(s+a)}\cdot\frac{1}{\Gamma(\frac{1}{2}s+\frac{1}{2}a)}\cdot\xi(s,\chi'), \tag{19}$$

where $a=0$ or 1, $b=0$ or 1, l is a non-negative integer, $f|k, p_\nu$ is a prime, ε_ν is a root of unity, and χ' is the proper character modulo f associated with χ.

(v) As in the case of $\xi(s)$, it can be shown that $\xi(s,\chi)$ is an entire function of order 1, and can therefore be factorized by Lemma C, Chapter II, and

$$L(s,\chi) = \frac{1}{s^b(s-1)^b} \cdot A e^{Bs} \cdot \frac{1}{\Gamma(\frac{1}{2}s+\frac{1}{2}a)} \cdot s^d \cdot \prod_\rho \left(1 - \frac{s}{\rho}\right) e^{s/\rho}, \tag{20}$$

where d is the number of ε_ν's such that $\varepsilon_\nu = 1$ (that is, the number of prime factors p of k, such that, in (1), $\chi'(p)=1$), and ρ runs through the zeros of $L(s,\chi)$, which are not $\leqslant 0$.

Since $\dfrac{1}{\Gamma(s)}$ is an entire function of order 1, (20) can be rewritten as

$$L(s,\chi) = \frac{1}{s^b(s-1)^b} \cdot \alpha e^{\beta s} s^{1-a+d} \prod_\lambda \left(1 - \frac{s}{\lambda}\right) e^{s/\lambda}, \tag{21}$$

where λ runs through all the zeros of $L(s,\chi)$ different from the origin. By logarithmic differentiation, one can deduce from this estimates for $\dfrac{L'(s,\chi)}{L(s,\chi)}$.

(vi) As in the case of $\zeta(s)$, one can deduce the existence of an infinity of non-real zeros of $L(s,\chi)$ either by an appeal to the theory of entire functions, or by an estimate of the Riemann-von Mangoldt type. One can prove that if $N(T)$ denotes the number of zeros $\rho = \beta + i\gamma$ of $L(s,\chi)$ for which $0 < \beta < 1$, and $0 < \gamma \leqslant T$, then

$$N(T) = \frac{1}{2\pi} T \log T + AT + O(\log T), \tag{22}$$

where A is a constant which depends on k, the modulus of the character χ.

(vii) The theorem on the existence of an infinity of zeros on the critical line, proved by Hardy in the case of $\zeta(s)$, has been extended to a general class of Dirichlet series, including the L-functions, by E. Hecke.

(viii) The analogue of Hamburger's theorem holds in a suitably modified form.

§ 5. Siegel's theorem. We need some further results on characters in preparation for the proof of Siegel's theorem (Theorem 3).

Let χ be any non-principal character modulo k. Then, for any positive integer n, we have

$$\left| \sum_{m=1}^{n} \chi(m) \right| \leqslant \tfrac{1}{2} k. \tag{23}$$

For let

$$\left.\begin{aligned} a_n &= \sum_{m=1}^{n} \chi(m), \qquad n=1,2,\dots, \\ a_0 &= 0. \end{aligned}\right\} \tag{24}$$

Then $|a_n - a_l| \leqslant |n-l|$, for $n \geqslant 0$, $l \geqslant 0$, and $a_l = 0$ if l is a multiple of k. Choosing $l = k\left[\frac{n}{k} + \frac{1}{2}\right]$, we get $|a_n| = |a_n - a_l| \leqslant \frac{1}{2}k$, since $l > k\left(\frac{n}{k} - \frac{1}{2}\right) = n - \frac{k}{2}$, and $l \leqslant k\left(\frac{n}{k} + \frac{1}{2}\right) = n + \frac{k}{2}$.

Using (23), we prove that

$$\left|\sum_{n=n_1}^{n_2} \chi(n) n^{-s}\right| \leqslant k \cdot \frac{|s|}{\sigma} \cdot n_1^{-\sigma}, \tag{25}$$

provided that $\sigma > 0$, $1 \leqslant n_1 \leqslant n_2$, and χ a non-principal character modulo k.

By partial summation, we have

$$\begin{aligned} \sum_{n=n_1}^{n_2} \chi(n) n^{-s} &= \sum_{n=n_1}^{n_2} (a_n - a_{n-1}) n^{-s} \\ &= \sum_{n=n_1}^{n_2} a_n (n^{-s} - (n+1)^{-s}) - a_{n_1 - 1} n_1^{-s} + a_{n_2}(n_2+1)^{-s}, \end{aligned}$$

and

$$|n^{-s} - (n+1)^{-s}| = \left|\int_n^{n+1} s x^{-s-1}\, dx\right| \leqslant |s| \int_n^{n+1} x^{-\sigma-1}\, dx.$$

Hence

$$\begin{aligned} \left|\sum_{n=n_1}^{n_2} \chi(n) n^{-s}\right| &\leqslant \tfrac{1}{2}k\left(|s| \int_{n_1}^{n_2+1} x^{-\sigma-1}\, dx + n_1^{-\sigma} + (n_2+1)^{-\sigma}\right) \\ &\leqslant \tfrac{1}{2}k\left(\frac{|s|}{\sigma}(n_1^{-\sigma} - (n_2+1)^{-\sigma}) + n_1^{-\sigma} + (n_2+1)^{-\sigma}\right) \\ &\leqslant k \cdot \frac{|s|}{\sigma} \cdot n_1^{-\sigma}, \end{aligned}$$

as claimed in (25).

It is immediate that $\sum_{n=1}^{\infty} \chi(n) n^{-s}$ converges uniformly on compact sets in the half-plane $\sigma > 0$, and therefore $L(s, \chi)$ is regular for $\sigma > 0$. If we take $n_1 = 1$, and let $n_2 \to \infty$, we get

$$|L(s, \chi)| \leqslant k \cdot \frac{|s|}{\sigma}, \qquad \sigma > 0. \tag{26}$$

In particular,

$$|L(1,\chi)| \leqslant k. \tag{27}$$

A better estimate, for $k \geqslant 3$, is given by

$$|L(1,\chi)| \leqslant 2+\log k. \tag{28}$$

If in (25) we choose $s=1$, $n_1=k$, and let $n_2 \to \infty$, we get

$$\left| \sum_{n=k}^{\infty} \chi(n) n^{-1} \right| \leqslant 1. \tag{29}$$

Hence

$$|L(1,\chi)| \leqslant \left| \sum_{n=1}^{k-1} \chi(n) n^{-1} \right| + 1 \leqslant 1 + \sum_{n=1}^{k-1} n^{-1} \leqslant 2 + \log k,$$

which proves (28).

If χ_0 is a real character modulo k_0, and χ a real character modulo k, then $\chi\chi_0$ is a real character modulo kk' (as noted in § 3). *There exists an arithmetical function* $c(n)$, *such that* $c(n) \geqslant 0$, $c(1)=1$, *and*

$$\zeta(s)\, L(s,\chi_0)\, L(s,\chi)\, L(s,\chi\chi_0) = \sum_{n=1}^{\infty} c(n) n^{-s}, \tag{30}$$

for $\sigma > 1$.

If p is a prime, we have $|\chi(p)p^{-s}| \leqslant p^{-\sigma} < 1$ for $\sigma > 1$, so that we define

$$\log\left(1 - \frac{\chi(p)}{p^s}\right)^{-1} = \sum_{l=1}^{\infty} \frac{\chi(p^l)}{l p^{ls}}.$$

Then the function $\log L(s,\chi)$ is uniquely defined for $\sigma > 1$, and given by

$$\log L(s,\chi) = \sum_{p,l} \frac{\chi(p^l)}{l p^{ls}},$$

where p runs through all the primes, and l through all positive integers. The double series is absolutely convergent for $\sigma > 1$. Further

$$e^{\log L(s,\chi)} = L(s,\chi).$$

Hence

$$\begin{aligned}
\log\zeta(s) + \log L(s,\chi_0) + \log L(s,\chi) + \log L(s,\chi\chi_0) &= \sum_{p,l} \frac{(1+\chi_0(p^l)+\chi(p^l)+\chi\chi_0(p^l))}{l p^{ls}} \\
&= \sum_{p,l} \frac{(1+\chi_0(p^l))(1+\chi(p^l))}{l p^{ls}} \\
&= \sum_{p,l} \frac{c(l,p)}{l p^{ls}},
\end{aligned}$$

where $c(l,p) \geqslant 0$. If we exponentiate this, we obtain (30).

If χ_0 modulo k_0, and χ modulo k, are any two equivalent non-principal characters, then $L(s,\chi_0)$ and $L(s,\chi)$ have the same zeros in the half-plane $\sigma>0$. (31)

To prove this, let

$$f_0(s)=\prod_{p\mid k_0 k}(1-\chi_0(p)p^{-s}),\qquad f(s)=\prod_{p\mid k_0 k}(1-\chi(p)p^{-s}). \tag{32}$$

For $\sigma>1$, we have

$$\begin{aligned} f_0(s)L(s,\chi_0) &= \prod_{p\nmid k_0 k}(1-\chi_0(p)p^{-s})^{-1} \\ &= \prod_{p\nmid k_0 k}(1-\chi(p)p^{-s})^{-1} = f(s)L(s,\chi). \end{aligned}$$

But $f_0(s)L(s,\chi_0)$ and $f(s)L(s,\chi)$ are regular for $\sigma>0$, hence

$$f_0(s)L(s,\chi_0)=f(s)L(s,\chi),\qquad \sigma>0. \tag{33}$$

Since the only possible zeros of f_0 and f are on the imaginary axis, assertion (31) follows.

If χ is a non-principal character modulo k, $k\geqslant 8$, and s is real, $s>1-\dfrac{1}{\log k}$, then

$$|L'(s,\chi)|<6\log^2 k, \tag{34}$$

where the dash indicates differentiation.

We have

$$\begin{aligned} L'(s,\chi) &= -\sum_{n=1}^{k}\chi(n)n^{-s}\log n - \sum_{n=k+1}^{\infty}\chi(n)n^{-s}\log n \\ &= -S_1-S_2, \quad \text{say,} \end{aligned}$$

where

$$\begin{aligned} |S_1| &\leqslant \sum_{n=1}^{k} n^{-1+\frac{1}{\log k}}\log n \leqslant k^{\frac{1}{\log k}}\cdot\log k\sum_{n=1}^{k}n^{-1} \\ &\leqslant e\log k(\log k+1)<\tfrac{3}{2}e\log^2 k, \end{aligned}$$

while

$$\begin{aligned} S_2 &= \sum_{n=k+1}^{\infty}(a_n-a_{n-1})n^{-s}\log n, \qquad a_n=\sum_{m=1}^{n}\chi(m), \\ &= \sum_{n=k+1}^{\infty}a_n(n^{-s}\log n-(n+1)^{-s}\log(n+1)). \end{aligned}$$

Since $n^{-s}\log n$ is a decreasing function of n for $n \geqslant k$, we have, by (23),

$$\begin{aligned}|S_2| &\leqslant \tfrac{1}{2}k \sum_{n=k+1}^{\infty} (n^{-s}\log n - (n+1)^{-s}\log(n+1)) \\ &= \tfrac{1}{2}k(k+1)^{-s}\log(k+1) \\ &< \tfrac{1}{2}k^{1-s}\log k < \tfrac{1}{2}e\log k < \tfrac{1}{2}e\log^2 k.\end{aligned}$$

Thus (34) is proved.

We are now in a position to prove the following crucial

LEMMA. *Let χ_0 modulo k_0, and χ modulo k, be two non-equivalent, non-principal, real characters, and let*

$$g(s) = \zeta(s)L(s,\chi_0)L(s,\chi)L(s,\chi_0\chi),$$
$$\rho = L(1,\chi_0)L(1,\chi)L(1,\chi_0\chi),$$

and

$$\tfrac{2}{3} < a < 1.$$

Then

$$g(a) > \frac{1}{2} - \frac{c_8\rho}{1-a}(k_0 k)^{c_6(1-a)}, \tag{35}$$

where c_6 and c_8 are absolute constants, and $c_6 > \frac{3}{2}$.

PROOF (ESTERMANN). By (30), we have

$$g(s) = \sum_{n=1}^{\infty} \frac{c(n)}{n^s}, \qquad \sigma > 1,$$

where $c(n) \geqslant 0$, $c(1) = 1$. Therefore $g(2) \geqslant 1$, and

$$(-1)^m g^{(m)}(2) = \sum_{n=1}^{\infty} \frac{c(n)}{n^2} \cdot \log^m n \geqslant 0, \quad \text{for} \quad m = 1,2,\ldots.$$

The Taylor expansion of g in a neighbourhood of the point $s=2$ is given by

$$g(s) = \sum_{m=0}^{\infty} a_m(2-s)^m, \qquad |s-2| < 1, \tag{36}$$

where $a_m \geqslant 0, a_0 \geqslant 1$. And $g(s) - \rho/(s-1)$ is an entire function. Hence

$$g(s) - \frac{\rho}{s-1} = \sum_{m=0}^{\infty} (a_m - \rho)(2-s)^m, \quad \text{for} \quad |s-2| < 2. \tag{37}$$

Using this, and an upper bound for $\left|g(s) - \dfrac{\rho}{s-1}\right|$ on the circle $|s-2| = \frac{3}{2}$, we shall derive an inequality for $|a_m - \rho|$.

For $|s-2|=\frac{3}{2}$, we have $\frac{\sigma^2}{|s|^2} \geqslant c_1 > 0$, so that, by (26),

$$|L(s,\chi_0)| \leqslant c_2 k_0, \qquad |L(s,\chi)| \leqslant c_2 k.$$

Since χ_0 and χ are real, and non-equivalent, $\chi_0 \chi$ is non-principal. For otherwise $\chi_0(n)=\chi(n)$ for $(n, k_0 k)=1$, which implies that χ_0 and χ may be extended to the same character modulo $k_0 k$. It follows that they are derived from the same proper character, which is impossible since they are non-equivalent (see § 3). Hence

$$|L(s,\chi_0 \chi)| \leqslant c_2 k_0 k.$$

Further, $\zeta(s)$ is bounded on the circle $|s-2|=\frac{3}{2}$. Hence

$$|g(s)| \leqslant c_3 k_0^2 k^2. \tag{38}$$

From (27) we also have $|\rho| \leqslant k_0^2 k^2$. Hence

$$\left| g(s) - \frac{\rho}{s-1} \right| < c_4 k_0^2 k^2, \qquad |s-2|=\tfrac{3}{2}.$$

By Cauchy's inequality, applied to (37), we get

$$|a_m - \rho| \leqslant c_4 k_0^2 k^2 (\tfrac{2}{3})^m, \qquad m=0,1,2,\ldots. \tag{39}$$

Since $0<a<1$, we have, by (37),

$$\begin{aligned} g(a) - \frac{\rho}{a-1} &= \sum_{m=0}^{\infty} (a_m - \rho)(2-a)^m \\ &= \left(\sum_{m=0}^{m_0-1} + \sum_{m=m_0}^{\infty} \right)(a_m-\rho)(2-a)^m, \qquad m_0 \geqslant 1, \\ &= A + B, \quad \text{say}. \end{aligned}$$

Since $a_0 \geqslant 1$, $a_m \geqslant 0$, $m=1,2,\ldots$, we have

$$A \geqslant 1 - \rho \sum_{m=0}^{m_0-1} (2-a)^m = 1 - \rho \frac{(2-a)^{m_0}-1}{1-a},$$

and, by (39),

$$B \geqslant -c_4 k_0^2 k^2 \sum_{m=m_0}^{\infty} (\tfrac{2}{3})^m (2-a)^m.$$

If $1>a>\frac{2}{3}$, then $\frac{2}{3}<\frac{2}{3}(2-a)=\eta<\frac{8}{9}$, so that

$$B \geqslant -c_4 k_0^2 k^2 \frac{\eta^{m_0}}{1-\eta} \geqslant -c_5 k_0^2 k^2 \eta^{m_0}, \qquad c_5 > 1.$$

Hence

$$g(a) \geqslant 1 - \frac{\rho(2-a)^{m_0}}{1-a} - c_5 k_0^2 k^2 \eta^{m_0}.$$

We now choose m_0, such that $c_5 k_c^2 k^2 \eta^{m_0} < \frac{1}{2}$, for example

$$m_0 = \left[\frac{\log(2 c_5 k_0^2 k^2)}{\log \frac{1}{\eta}} \right] + 1.$$

Then $m_0 \geqslant 1$, and $m_0 < c_6 \log(k_0 k) + c_7$, where $c_6 > \frac{3}{2}$, $c_7 > 1$. Hence

$$(2-a)^{m_0} < (\tfrac{3}{2}\eta)^{c_7} (k_0 k)^{c_6 \log(2-a)} < c_8 (k_0 k)^{c_6(1-a)}.$$

Thus, if $\frac{2}{3} < a < 1$, then

$$g(a) > \frac{1}{2} - \frac{c_8 \rho}{1-a} (k_0 k)^{c_6(1-a)},$$

where c_6 and c_8 are positive, absolute constants, and $c_6 > \frac{3}{2}$.

THEOREM 3 (SIEGEL). *Given any $\varepsilon > 0$, there exists a number $A = A(\varepsilon) > 0$, such that if $k > A$, and χ a real, non-principal character modulo k, then $L(s, \chi)$ has no zeros on the real axis between $1 - k^{-\varepsilon}$ and 1.*

PROOF. There is no loss of generality in supposing that $0 < \varepsilon < 1$. If no L-function corresponding to a real, non-principal character modulo k, has a zero on the real axis between $1 - \dfrac{\varepsilon}{2 c_6}$ and 1, where c_6 is the constant in (35), then the result is trivial, for we have only to take $A = \left(\dfrac{\varepsilon}{2 c_6}\right)^{-1/\varepsilon}$, in which case $1 - k^{-\varepsilon} > 1 - \dfrac{\varepsilon}{2 c_6}$.

We therefore assume that there exists an integer k_0, a real, non-principal character χ_0 modulo k_0, and a number a, such that

$$1 - \frac{\varepsilon}{2 c_6} < a < 1,$$

and

$$L(a, \chi_0) = 0. \tag{40}$$

Since $L(s, \chi_0)$ cannot have an infinity of zeros between a and 1, it follows that there exists a number a_1, such that $a \leqslant a_1 < 1$, and

$$L(s, \chi_0) \neq 0, \quad \text{for} \quad a_1 < s < 1. \tag{41}$$

If χ is any real, non-principal character modulo k, which is equivalent to χ_0, then, by (31), $L(s, \chi_0)$ and $L(s, \chi)$ have the same zeros in the half-plane $\sigma>0$, and if $k>(1-a_1)^{-1/\varepsilon}$, it follows that $1-k^{-\varepsilon}>a_1$, hence

$$L(s,\chi)\neq 0, \quad \text{for} \quad 1-k^{-\varepsilon}<s<1,$$

as claimed in the theorem, with $A=(1-a_1)^{-1/\varepsilon}$.

If χ is any real, non-principal character, which is *not* equivalent to χ_0, then we appeal to the lemma proved above. Noting (40), and the fact that the a chosen satisfies $1>a>\frac{2}{3}$, since $c_6>\frac{3}{2}$ and $0<\varepsilon<1$, we have

$$g(a)=0>\frac{1}{2}-\frac{c_8}{1-a}L(1,\chi_0)\,L(1,\chi)\,L(1,\chi_0\chi)\,(k_0k)^{c_6(1-a)},$$

where $\chi_0\chi$ is real and non-principal since χ_0 and χ are real and non-equivalent. By Euler's identity, $L(1,\chi_0)\geqslant 0$, $L(1,\chi)\geqslant 0$, while by (28), $L(1,\chi\chi_0)\leqslant 2+\log(k_0k)$. Hence

$$\frac{c_9}{1-a}L(1,\chi_0)\,L(1,\chi)\,(2+\log k_0k)\,(k_0k)^{c_6(1-a)}>1, \qquad c_9=2c_8.$$

Now choose $A>8$, such that for $k>A$, we have

$$\log k<k^{\varepsilon},$$

and

$$\frac{k^{\varepsilon/2}}{6\log^2 k}>\frac{c_9}{1-a}L(1,\chi_0)\,(2+\log(k_0k))\,k_0^{c_6(1-a)}.$$

Then for $k>A$, we have

$$L(1,\chi)\,k^{c_6(1-a)+\frac{1}{2}\varepsilon}>6\log^2 k.$$

Since $1-\dfrac{\varepsilon}{2c_6}<a<1$, it follows that

$$L(1,\chi)>6\log^2 k\cdot k^{-(\frac{1}{2}\varepsilon+\frac{1}{2}\varepsilon)}=6\log^2 k\cdot k^{-\varepsilon}. \tag{42}$$

If there exists an a_2, such that $1-k^{-\varepsilon}<a_2<1$, for which $L(a_2,\chi)=0$, then by the mean-value theorem of differential calculus, there would also exist an a_3, such that $a_3>1-k^{-\varepsilon}$, for which

$$L(1,\chi)=(1-a_2)\,L'(a_3,\chi).$$

Since $\log k<k^{\varepsilon}$ for $k>A$, we have $a_3>1-\dfrac{1}{\log k}$. Hence, by (34), we have $L'(a_3,\chi)<6\log^2 k$ (since $k\geqslant 8$). Thus

$$L(1,\chi)<6\,k^{-\varepsilon}\log^2 k, \tag{43}$$

which contradicts (42). Hence $L(s,\chi)$ is not zero for $1-k^{-\varepsilon}<s<1$, and the theorem follows.

REMARKS.

(i) Let $\left(\frac{d}{n}\right)$ denote Kronecker's extension of the Legendre symbol for quadratic residues, for $n=1,2,\ldots$.

It is known that if d is the discriminant of a *quadratic number field* (that is, d is square-free and $\equiv 1 \pmod 4$; or $d=4N$, where $N\equiv 2$ or $3 \pmod 4$, and square-free), and $\chi(n)=\left(\frac{d}{n}\right)$, then χ is a real, proper character modulo $|d|$. Conversely if χ is a real, proper character modulo $|d|$, $d\neq 0$, then $\chi(n)=\left(\frac{d}{n}\right)$, $n=1,2,\ldots$, where d is the discriminant of a quadratic number field.

Thus $L(s,\chi)=\sum_{n=1}^{\infty}\left(\frac{d}{n}\right)n^{-s}$, for $\sigma>1$, for any real, proper character χ modulo $|d|$, where d is a suitably chosen integer different from zero.

The case $d=1$, corresponding to the rational number field, and to the proper character $\chi\equiv 1$, gives rise to the Riemann zeta-function.

(ii) The proof of Theorem 3 is actually contained in Siegel's proof of the result

$$L(1,\chi)=o(\log|d|), \quad \text{as} \quad |d|\to\infty, \tag{44}$$

where d is the discriminant of a quadratic field, and $\chi(n)=\left(\frac{d}{n}\right)$.

Since it is known, after Dirichlet, that the *class number* $h(d)$ of a quadratic field is given by

$$\pi(|d|)^{-\frac{1}{2}}h(d)=L(1,\chi), \quad \text{if} \quad d<-4,$$

and

$$2d^{-\frac{1}{2}}h(d)\log\varepsilon_d=L(1,\chi), \quad \text{if} \quad d>0,$$

where ε_d is the fundamental unit, (44) is equivalent to the assertion that

$$\log h(d)\sim\log\sqrt{|d|}, \quad \text{as} \quad d\to-\infty,$$

and

$$\log(h(d)\log\varepsilon_d)\sim\log\sqrt{d}, \quad \text{as} \quad d\to+\infty. \tag{45}$$

This includes the important result previously obtained by Heilbronn that $h(d)\to\infty$ as $d\to-\infty$, in confirmation of a conjecture of Gauss.

(iii) By an application of Theorem 3, Walfisz deduced that

$$\pi(x;k,l)-\frac{\operatorname{li}x}{\varphi(k)}=O(xe^{-c\sqrt{\log x}}), \tag{46}$$

uniformly for $3\leqslant k\leqslant\log^{\alpha}x$, where α is any positive number however large, and c a positive, absolute constant.

(iv) An estimate of the type (46) is of crucial importance in the proof of Vinogradov's important theorem that every sufficiently large odd number is a sum of three primes. Instead of (46) one can directly use Theorem 3, as Chudakov has done. The connexion between the location of the zeros of L-functions and the problem of expressing every sufficiently large odd number as a sum of three primes was first realized by Hardy and Littlewood.

Notes on Chapter VI

As general references, see C. L. Siegel, *Acta Arithmetica*, 1 (1935), 83—86; *Gesammelte Abhandlungen*, Bd. 1, 21; T. Estermann, *Introduction to modern prime number theory*, Cambridge Tract, No. 41 (1952), Theorem 48; H. Davenport, *Multiplicative number theory*, (1967), § 21.

§ 1. For an elementary exposition of characters and L-functions, and for a proof of Dirichlet's theorem on the existence of an infinity of primes in every arithmetical progression, see, for instance, the author's *Introduction*, Ch. X.

§ 2. For Walfisz's application of Siegel's theorem to the estimation of $\pi(x;k,l)$, see *Math. Zeitschrift*, 40 (1936), 593, Hilfssatz 3. For Landau's approximation of $\pi(x;k,l)$ see his *Vorlesungen*, II, Satz 403. Chudakov's theorem is proved in *Mat. Sbornik*, (1) 43 (1936), 591—602.

§§ 3—4. For the functional equation of $L(s,\chi)$, see, for instance, Landau's *Primzahlen*, §§ 126—128, or Davenport, loc. cit., § 9. For Remarks (i)—(vi) see Landau's *Primzahlen*, §§ 129—140. For Remark (vii) see E. Hecke, *München Akad. Sitzungsberichte*, II, 8 (1937), 73—95; *Werke*, 708. For Remark (viii) see S. Bochner and K. Chandrasekharan, *Annals of Math.* 63 (1956), 353, § 7; H. Hamburger, *Math. Zeitschrift*, 10 (1921), 240—254; 11 (1922), 224—245; 13 (1922), 283—311.

§ 5. The proof of Siegel's theorem given here is the version formulated by T. Estermann, *J. London Math. Soc.* 23 (1948), 275—279. Another version has been given by S. Chowla, *Annals of Math.* (2) 51 (1950), 120—122. They make no use of algebraic number theory. But the underlying idea of both the proofs is not different from Siegel's own (loc. cit) which is a model of style and elegance. Siegel himself has given a proof of his theorem without the use of algebraic number theory. See Landau, *Über einige neuere Fortschritte der additiven Zahlentheorie*, Cambridge Tract, No. 35 (1937), 85.

For Remarks (i) and (ii) see Davenport, loc. cit., 37—44; 130—144; Siegel, loc. cit., and H. Heilbronn, *Quarterly J. of Math.* (Oxford) 5 (1934), 150—160. For the properties of the Kronecker symbol see Landau's *Vorlesungen*, I, Kap. 6.

Siegel's conjecture that his *class-number formula* (45) could be extended to algebraic number fields of arbitrary degree was proved by R. Brauer, *American J. Math.* 69 (1947), 243—250.

For Remark (iii) see Walfisz, loc. cit. in § 2, as well as Estermann loc. cit. at the beginning of these notes, Theorem 55.

For Remark (iv) see I. M. Vinogradov, *The method of trigonometrical sums in the theory of numbers*, Interscience, New York (1955), Ch. X, and, in particular, the notes on that chapter. See also Chudakov (=Tchudakoff), *Annals of Math.* 48 (1947), 515—545 (see p. 540); G. H. Hardy and J. E. Littlewood, *Acta Math.* 44 (1923) 1—70.

E. Bombieri has proved that if $\sum_{n=1}^{m} \chi(n) \geqslant -q$, for all m, then $L(1,\chi) \geqslant \dfrac{\pi}{16(q+1)}$, *Rendiconti Istituto Lombardo Accad. Sc. Lettere* (A) (1960), 642—649.

Chapter VII

Theorems of Hardy-Ramanujan and of Rademacher on the partition function

§ 1. The partition function. In the preceding chapters we have considered the asymptotic behaviour of arithmetical functions connected, in one way or another, with the distribution of primes. We shall now consider the *partition function* which arises from an altogether different context.

An *unrestricted partition*, or merely a *partition*, of a positive integer n is a representation of n as a sum of strictly positive integers. Two representations which differ only in the order of their summands are considered identical. Let $p(n)$ denote the number of partitions of the positive integer n. Thus $p(1)=1$, $p(4)=5$, and $p(5)=7$. We define $p(0)=1$. The map $n \to p(n)$ defines the partition function.

Our object is to estimate the order of magnitude of $p(n)$ as $n \to \infty$. We shall prove the following asymptotic formula due to G. H. Hardy and S. Ramanujan:

$$p(n) = \frac{e^{K\lambda_n}}{4\sqrt{3}\cdot\lambda_n^2} + O\left(\frac{e^{K\lambda_n}}{\lambda_n^3}\right), \quad K=\pi\sqrt{\tfrac{2}{3}}, \quad \lambda_n=\sqrt{n-\tfrac{1}{24}}, \quad n \geqslant 1. \tag{1}$$

We shall also establish the following identity due to H. Rademacher:

$$p(n) = \frac{1}{\pi\sqrt{2}} \sum_{q=1}^{\infty} A_q(n)\, q^{\frac{1}{2}} \frac{d}{dn}\left(\frac{\sinh\left(\frac{K}{q}\sqrt{n-\frac{1}{24}}\right)}{\sqrt{n-\frac{1}{24}}}\right), \quad n \geqslant 1, \tag{2}$$

where $A_q(n)$ is a sum of roots of unity.

We shall follow C. L. Siegel in our presentation, in view of the many simplifications which he has effected in the proof of (1) and (2).

§ 2. A simple case. To illustrate the nature of the general problem, we consider the simple case of finding the number of representations $a(n)$ of the positive integer n as a sum of *three* non-negative integers, the order of the summands being considered irrelevant, and repetition of the summands being allowed. We define $a(0)=1$. Clearly $a(n)$ equals the number of solutions of the equation

$$n=x_1+x_2+x_3, \tag{3}$$

in non-negative integers x_1, x_2, x_3. Since the order of the summands is irrelevant, we may assume that $x_1 \geqslant x_2 \geqslant x_3$. If we set $x_1-x_2=y_1$, $x_2-x_3=y_2$, and $x_3=y_3$, then equation (3) becomes

$$n=y_1+2y_2+3y_3, \quad y_k \geqslant 0, \quad k=1,2,3. \tag{4}$$

The number of solutions of (4) is the same as that of (3), and equals $a(n)$. The *generating function* of $a(n)$ is defined by

$$f(t)=\sum_{n=0}^{\infty} a(n)t^n, \quad |t|<1. \tag{5}$$

It is easy to see that f is a rational function, since

$$f(t)=\sum_{y_1=0}^{\infty} t^{y_1} \cdot \sum_{y_2=0}^{\infty} t^{2y_2} \cdot \sum_{y_3=0}^{\infty} t^{3y_3} = \frac{1}{(1-t)} \cdot \frac{1}{(1-t^2)} \cdot \frac{1}{(1-t^3)}.$$

We can express f as a sum of partial fractions, namely

$$f(t)=\frac{1}{6(1-t)^3}+\frac{1}{4(1-t)^2}+\frac{17}{72(1-t)}+\frac{1}{8(1+t)}+\frac{1}{9(1-\omega t)}+\frac{1}{9(1-\omega^2 t)}, \tag{6}$$

where ω and ω^2 denote the two complex cube roots of unity. Expanding each of the fractions in a power series, and picking out the coefficient of t^n, we obtain, from (5) and (6),

$$\begin{aligned} a(n) &= \frac{1}{6}\,\frac{(n+1)(n+2)}{2}+\frac{1}{4}(n+1)+\frac{17}{72}+\frac{1}{8}(-1)^n+\frac{1}{9}(\omega^n+\omega^{2n}) \\ &= \frac{(n+3)^2}{12}-\frac{7}{72}+\frac{1}{8}(-1)^n+\frac{1}{9}(\omega^n+\omega^{2n}). \end{aligned}$$

If we set

$$b(n)=a(n)-\frac{(n+3)^2}{12}, \tag{7}$$

then

$$|b(n)| \leqslant \tfrac{7}{72}+\tfrac{1}{8}+\tfrac{2}{9}=\tfrac{32}{72}<\tfrac{1}{2}.$$

Since $a(n)$ must be an integer, it follows that *$a(n)$ is the integer nearest to* $\dfrac{(n+3)^2}{12}$.

If $n=9$, for example, this formula gives $a(9)=12$, which can be verified directly.

Imitating this procedure, we find that $p(n)$ is the number of solutions of the equation

$$n=y_1+2y_2+3y_3+\cdots, \qquad y_k\geqslant 0, k=1,2,\ldots, \tag{8}$$

where y_1 denotes the number of 1's in the partition of n, y_2 the number of 2's, and so on, all but a finite number of the y's being zero. If $f(t)=\prod_{k=1}^{\infty}(1-t^k)^{-1}$, $|t|<1$, we shall see that $f(t)$ is the *generating function* of $p(n)$, namely

$$f(t)=\prod_{k=1}^{\infty}(1-t^k)^{-1}=\sum_{n=0}^{\infty}p(n)t^n, \qquad |t|<1. \tag{9}$$

For let $0\leqslant t<1$, and m an integer >1, and

$$f_m(t)=\prod_{k=1}^{m}(1-t^k)^{-1}=1+\sum_{n=1}^{\infty}p_m(n)t^n.$$

Then $p_m(n)\leqslant p(n)$, $p_m(n)=p(n)$ for $1\leqslant n\leqslant m$, and $p_m(n)\to p(n)$ as $m\to\infty$, for every n. Therefore

$$f_m(t)=1+\sum_{n=1}^{m}p(n)t^n+\sum_{n=m+1}^{\infty}p_m(n)t^n.$$

Now $f_m(t)\leqslant f(t)$ for $0\leqslant t<1$, and $f_m(t)\to f(t)$ as $m\to\infty$. Hence

$$1+\sum_{n=1}^{m}p(n)t^n\leqslant f_m(t)\leqslant f(t),$$

so that $\sum_{n=1}^{\infty}p(n)t^n$ converges, hence also $\sum_{n=1}^{\infty}p_m(n)t^n$ for any fixed t such that $0\leqslant t<1$, uniformly for all values of m. It follows that

$$1+\sum_{n=1}^{\infty}p(n)t^n=\lim_{m\to\infty}\left(1+\sum_{n=1}^{\infty}p_m(n)t^n\right)=\lim_{m\to\infty}f_m(t)=f(t),$$

for $0\leqslant t<1$. The extension of the result for $|t|<1$ is immediate, and (9) is proved.

The method of partial fractions does not work in the general case, but Cauchy's integral formula gives

$$p(n)=\frac{1}{2\pi i}\int_C\frac{f(t)\,dt}{t^{n+1}}, \tag{10}$$

where C is a suitable circle enclosing the origin in the complex t-plane. Since it turns out that the unit circle is a natural boundary for the function f, we choose a circle C whose radius is less than 1, depends on n, and tends to 1 as $n\to\infty$. This method of finding $p(n)$ requires considerable care in execution, and leads to the asymptotic formula (1) originally given by Hardy and Ramanujan. By a refinement of their argument, the convergent series for $p(n)$, given in (2), can be obtained, as was originally done by Rademacher.

§ 3. A bound for $p(n)$. Compared with the asymptotic formula (1), it is easy to obtain the inequality

$$p(n) < e^{K\sqrt{n}}, \qquad K = \pi\sqrt{\tfrac{2}{3}}. \tag{11}$$

If $0<t<1$, and we take logarithms in (9), we get

$$g(t) = \log f(t) = \sum_{k=1}^{\infty} \log\left(\frac{1}{1-t^k}\right) = \sum_{k=1}^{\infty}\sum_{j=1}^{\infty} \frac{t^{kj}}{j} = \sum_{j=1}^{\infty} \frac{1}{j}\,\frac{t^j}{1-t^j}, \tag{12}$$

the interchange in the order of summation being permitted since all the terms are positive. Now, if j is an integer, and $j\geqslant 1$, and $0<t<1$, then

$$\frac{1-t^j}{1-t} = 1+t+t^2+\cdots+t^{j-1},$$

so that

$$j t^{j-1} < \frac{1-t^j}{1-t} < j, \qquad 0<t<1, \qquad 1\leqslant j, \quad j \text{ integral}.$$

If we use this in (12), we obtain

$$\sum_{j=1}^{\infty} \frac{t^{j-1}}{j^2} < \frac{(1-t)}{t}\, g(t) < \sum_{j=1}^{\infty} \frac{1}{j^2}. \tag{13}$$

Letting $t\to 1-0$, we obtain

$$\lim_{t\to 1-0} \frac{(1-t)g(t)}{t} = \frac{\pi^2}{6}, \tag{14}$$

since $\sum_{j=1}^{\infty} \frac{1}{j^2} = \frac{\pi^2}{6}$. If $0<t<1$, all the terms in the expansion of $f(t)$ are positive, so that $p(n)t^n < f(t)$, for $n\geqslant 0$, which implies that

$$\log p(n) + n\log t < g(t) < \frac{\pi^2}{6}\cdot\frac{t}{1-t},$$

by (13). Thus

$$\log p(n) < \frac{\pi^2}{6}\frac{t}{(1-t)} + n\log\frac{1}{t}.$$

If we set $1+u=\frac{1}{t}$, then $0<u<\infty$ for $0<t<1$, and since $\log(1+u)<u$ for $u>0$, we obtain

$$\log p(n) < \frac{\pi^2}{6}\frac{1}{u} + nu.$$

The minimum of the right-hand side, as a function of u, is attained for $\frac{\pi^2}{6}\cdot\frac{1}{u}=nu$, or for $u=\frac{\pi}{\sqrt{6n}}$. Hence

$$\log p(n) < \frac{2\pi n}{\sqrt{6n}} = K\sqrt{n},$$

which proves (11).

§ 4. A property of the generating function of $p(n)$. Formula (9) gives the generating function of $p(n)$, namely

$$f(t) = \sum_{n=0}^{\infty} p(n)t^n = \prod_{k=1}^{\infty}\frac{1}{(1-t^k)}, \qquad 0<t<1.$$

Let $t=e^{-2\pi x}$, so that $x>0$, and let

$$f(t)=F(x) = \prod_{k=1}^{\infty}(1-e^{-2\pi kx})^{-1}. \tag{15}$$

If $\log f(t)=g(t)=G(x)$, then by (12) we have

$$G(x) = \sum_{k=1}^{\infty}\frac{1}{k(e^{2\pi kx}-1)}. \tag{16}$$

We shall show that F has the property

$$e^{\frac{\pi x}{12}}F(x)=\sqrt{x}\cdot e^{\frac{\pi}{12x}}F\left(\frac{1}{x}\right), \qquad x>0, \tag{17}$$

which is equivalent to

$$G(x) + \frac{\pi x}{12} = \log(\sqrt{x}) + \frac{\pi}{12x} + G\left(\frac{1}{x}\right), \qquad x>0. \tag{18}$$

This implies asymptotic formula (14). For $x = -\dfrac{\log t}{2\pi}$, so that $x \sim \dfrac{1-t}{2\pi}$ as $t \to 1-0$, and since $G\left(\dfrac{1}{x}\right) \to 0$ as $x \to 0$, formula (18) gives

$$g(t) = G(x) \sim \frac{\pi}{12x} \sim \frac{\pi^2}{6(1-t)}, \quad \text{as} \quad t \to 1,$$

which is equivalent to (14).

Since

$$\pi \cot \pi y = \frac{\pi i (e^{2\pi i y}+1)}{e^{2\pi i y}-1} = \pi i + \frac{2\pi i}{e^{2\pi i y}-1},$$

and

$$\pi \cot \pi y = \lim_{n\to\infty} \sum_{j=-n}^{n} \frac{1}{y-j} = \frac{1}{y} + \sum_{j=1}^{\infty} \frac{2y}{y^2-j^2},$$

we obtain by subtracting the second equation from the first, and dividing by $2\pi i$,

$$\frac{1}{e^{2\pi i y}-1} = -\frac{1}{2} + \frac{1}{2\pi i y} + \frac{1}{\pi} \sum_{j=1}^{\infty} \frac{iy}{j^2-y^2}.$$

If we put $iy = kx$, we get

$$\frac{1}{e^{2\pi k x}-1} = -\frac{1}{2} + \frac{1}{2\pi k x} + \frac{1}{\pi} \sum_{j=1}^{\infty} \frac{kx}{j^2+k^2x^2},$$

and if we substitute this in (16), we obtain

$$G(x) = \sum_{k=1}^{\infty} \left(-\frac{1}{2k} + \frac{1}{2\pi k^2 x} + \frac{1}{\pi} \sum_{j=1}^{\infty} \frac{1}{(j^2 x^{-1} + k^2 x)} \right),$$

or

$$\pi G(x) - \frac{\pi^2}{12x} = \sum_{k=1}^{\infty} \left(-\frac{\pi}{2k} + \sum_{j=1}^{\infty} \frac{1}{(j^2 x^{-1} + k^2 x)} \right), \quad \text{since} \quad \sum_{k=1}^{\infty} \frac{1}{k^2} = \frac{\pi^2}{6},$$

$$= \lim_{n\to\infty} \sum_{k=1}^{n} \left[-\frac{\pi}{2k} + \left(\sum_{j=1}^{n} + \sum_{n+1}^{\infty} \right) \frac{1}{(j^2 x^{-1} + k^2 x)} \right]. \tag{19}$$

If $u > 0$, $x > 0$, the function $(u + v^2 x^{-1})^{-1}$ decreases as $v (> 0)$ increases (for fixed u and x), so that

$$\left(u + \frac{(j+1)^2}{x} \right)^{-1} < \int_j^{j+1} \left(u + \frac{v^2}{x} \right)^{-1} dv < \left(u + \frac{j^2}{x} \right)^{-1}.$$

Hence

$$\sum_{j=n+1}^{\infty}\left(u+\frac{j^2}{x}\right)^{-1}<\int_n^{\infty}\left(u+\frac{v^2}{x}\right)^{-1}dv<\sum_{j=n}^{\infty}\left(u+\frac{j^2}{x}\right)^{-1},$$

which implies that

$$0<\int_n^{\infty}\left(u+\frac{v^2}{x}\right)^{-1}dv-\sum_{j=n+1}^{\infty}\left(u+\frac{j^2}{x}\right)^{-1}<\left(u+\frac{n^2}{x}\right)^{-1}<\frac{x}{n^2}, \tag{20}$$

if $u>0, x>0$.

Taking $u=k^2x$, and using (20) in (19), we obtain

$$\pi G(x)-\frac{\pi^2}{12x}=\lim_{n\to\infty}\sum_{k=1}^{n}\left[-\frac{\pi}{2k}+\sum_{j=1}^{n}\left(k^2x+\frac{j^2}{x}\right)^{-1}+\int_n^{\infty}\left(k^2x+\frac{v^2}{x}\right)^{-1}dv\right], \tag{21}$$

since the error which arises from the replacement of the sum $\sum\limits_{n+1}^{\infty}$ by the integral $\int\limits_n^{\infty}$ is $O\left(\frac{x}{n^2}\right)$ uniformly in k, where k runs from 1 to n, so that the total error is $O\left(\frac{x}{n}\right)=o(1)$, as $n\to\infty$.

By making the substitution $v=nxw$, we write (21) as

$$\pi G(x)-\frac{\pi^2}{12x}$$

$$=\lim_{n\to\infty}\left(\sum_{k=1}^{n}\left(-\frac{\pi}{2k}\right)+\sum_{k=1}^{n}\sum_{j=1}^{n}(k^2x+j^2x^{-1})^{-1}+\sum_{k=1}^{n}\int_{1/x}^{\infty}\frac{1}{n(w^2+k^2n^{-2})}dw\right).$$

If we replace x by $\frac{1}{x}$, and subtract the resulting equation from this, we get

$$\pi G(x)-\frac{\pi^2}{12x}-\pi G\left(\frac{1}{x}\right)+\frac{\pi^2x}{12}=\lim_{n\to\infty}\int_{1/x}^{x}\sum_{k=1}^{n}\frac{1}{n(w^2+k^2n^{-2})}dw=I,\text{ say.} \tag{22}$$

Since the integrand on the right-hand side is positive, we may take the limit under the integral sign. Further, by the definition of the Riemann integral, we have

$$\int_0^1 (u^2+w^2)^{-1}\,du = \lim_{n\to\infty} \sum_{k=1}^{n} \left(\left(\frac{k}{n}\right)^2 + w^2\right)^{-1} \cdot \frac{1}{n}.$$

Hence

$$I = \int_{1/x}^{x} dw \int_0^1 (u^2+w^2)^{-1}\,du.$$

Replacing u by $\frac{1}{u}$, and w by $\frac{1}{w}$, we obtain

$$I = \int_{1/x}^{x} dw \int_1^{\infty} (u^2+w^2)^{-1}\,du,$$

so that

$$I = \frac{1}{2}\int_{1/x}^{x} dw \int_0^{\infty} (u^2+w^2)^{-1}\,du = \frac{\pi}{4}\int_{1/x}^{x} \frac{dw}{w} = \frac{\pi}{2}\log x.$$

Using this in (22), we get

$$G(x) + \frac{\pi x}{12} = G\left(\frac{1}{x}\right) + \frac{\pi}{12x} + \log\sqrt{x}, \qquad x>0,$$

which is (18), and (17) is an immediate consequence. Since both sides of (17) are analytic functions of the complex variable x, (17) holds, by analytic continuation, for complex x with positive real part, and so does (18). Thus we have

THEOREM 1. *If*

$$F(x) = \prod_{k=1}^{\infty} (1-e^{-2\pi kx})^{-1}, \qquad \operatorname{Re} x>0, \tag{23}$$

then

$$e^{\frac{\pi x}{12}} F(x) = \sqrt{x}\cdot e^{\frac{\pi}{12x}} F\left(\frac{1}{x}\right), \qquad \operatorname{Re} x>0, \tag{24}$$

where that branch of the square-root is taken which is positive for positive x.

Equivalently, if

$$G(x) = \sum_{k=1}^{\infty} k^{-1}(e^{2\pi kx}-1)^{-1}, \qquad \operatorname{Re} x>0, \tag{25}$$

then

$$G(x) + \frac{\pi x}{12} = \log(\sqrt{x}) + \frac{\pi}{12x} + G\left(\frac{1}{x}\right), \qquad \operatorname{Re} x>0. \tag{26}$$

§ 5. The Dedekind η-function. Formula (24) is related to the transformation formula for the Dedekind η-function, which is defined by

$$\eta(z)=e^{\frac{\pi i z}{12}} \prod_{n=1}^{\infty} (1-e^{2\pi i n z}), \tag{27}$$

where z is a complex variable, with $\operatorname{Im} z>0$. The infinite product in (27) is absolutely convergent for $\operatorname{Im} z>0$, with non-zero factors, hence $\eta(z)\neq 0$ for $\operatorname{Im} z>0$. It converges uniformly in every compact subset of the half-plane $\operatorname{Im} z>0$, and therefore defines a regular analytic function of z for $\operatorname{Im} z>0$. Further

$$\eta(z\pm 1)=e^{\pm\frac{\pi i}{12}}\eta(z). \tag{28}$$

By (24) we have

$$\sqrt{u}\cdot e^{-\frac{\pi u}{12}} \prod_{k=1}^{\infty} (1-e^{-2\pi k u})=e^{-\frac{\pi}{12u}} \prod_{k=1}^{\infty} \left(1-e^{-\frac{2\pi k}{u}}\right), \tag{29}$$

for $u>0$, where that branch of the square-root is taken which is positive for positive u. As already remarked, this holds also for complex u with positive real part. If $u=-iz$, with $z=x+iy$, we have, for $y>0$,

$$\sqrt{\frac{z}{i}}\,\eta(z)=\eta\left(-\frac{1}{z}\right),$$

that branch of the square-root being taken which is positive for $z=i$, that is

$$e^{-\frac{\pi i}{4}}\cdot\sqrt{z}\cdot\eta(z)=\eta\left(-\frac{1}{z}\right), \quad \text{with} \quad \sqrt{z}=e^{\frac{\pi i}{4}} \quad \text{for} \quad z=i. \tag{30}$$

We shall prove a formula connecting $\eta(z)$ with $\eta\left(\frac{az+b}{cz+d}\right)$, where a, b, c, d are integers such that $ad-bc=1$, which gives (28) in case $a=1$, $b=\pm 1$, $c=0$, $d=1$, and (30) in case $a=0$, $b=-1$, $c=1$, $d=0$. For the proof of that formula we need some preliminaries.

Let a, b, c, d be integers, with $ad-bc=1$. The transformation M, defined by

$$z \to z' = \frac{az+b}{cz+d},$$

is called a *modular transformation.* It transforms the upper half-plane $\operatorname{Im} z>0$ into itself. For if $z=x+iy$, and $z'=x'+iy'$, then

$$\begin{aligned} y' &= \frac{1}{2i}(z'-\bar{z}') = \frac{1}{2i}\left(\frac{az+b}{cz+d} - \frac{a\bar{z}+b}{c\bar{z}+d}\right) \\ &= \frac{1}{2i}(z-\bar{z})|cz+d|^{-2} = y|cz+d|^{-2} > 0. \end{aligned}$$

If $z \to z' = \dfrac{az+b}{cz+d}$ is a modular transformation M, say, and $z' \to z'' = \dfrac{a'z'+b'}{c'z'+d'}$ another modular transformation M', say, then $z \to z''$ defines a modular transformation M'', called the *product* of M and M', and written $M''=M'M$, since $z'' = \dfrac{\alpha z+\beta}{\gamma z+\delta}$, where α, β, γ, δ are integers with the property $\alpha\delta-\beta\gamma=1$. The product is *associative*. The *inverse* of the modular transformation M is the modular transformation M^{-1} defined by

$$z \to \frac{dz-b}{-cz+a},$$

for if $Mz=z' = \dfrac{az+b}{cz+d}$, then $M^{-1}Mz = M^{-1}z' = \dfrac{dz'-b}{-cz'+a} = z = MM^{-1}z$. If $a=d=1$, and $b=c=0$, the transformation M reduces to the *identity* transformation.

We note that $-a$, $-b$, $-c$, $-d$, and a, b, c, d correspond to the same M, so that we may assume that $c \geqslant 0$, and that $d>0$ if $c=0$.

It follows that *the set of all modular transformations is a group. This group is generated by the two elements*

$$A\colon z \to -\frac{1}{z},$$

and (31)

$$B\colon z \to z+1.$$

To prove this, we have to prove that the general modular transformation M can be expressed as a product of powers of A and B.

Let $m=\min(|c|,|d|)$. If $|c|>|d|$ in M, then the transformation MA, defined by $z\to\dfrac{a'z+b'}{c'z+d'}$, has the property $|c'|<|d'|$, for

$$MAz=\frac{-\dfrac{a}{z}+b}{-\dfrac{c}{z}+d}=\frac{bz-a}{dz-c}.$$

Thus it is sufficient to consider a modular transformation M subject to the restriction $|c|\leqslant|d|$.

Let $|c|\leqslant|d|$, and let $m=0$. Then $c=0$, and $ad=1$. Hence M is given by $z\to z\pm b$, so that $M=B^{\pm b}$.

Let us now assume that M is generated by A and B for $m=0,1,2,\ldots,n-1$, where n is a positive integer. We shall prove that it is then so for $m=n$.

By the definition of m, and by the assumption $|c|\leqslant|d|$, we have $m=|c|$. We may assume $c>0$, for otherwise we can multiply a, b, c, d by -1, which leaves M unchanged. Then $m=c$, and we have to prove that M is generated by A and B also for $c=n$.

Consider the transformation MB^k given by

$$z\to\frac{az+(ak+b)}{cz+(ck+d)},$$

where k is an integer such that $0\leqslant ck+d<c$. For that choice of k, we have $\min(c,ck+d)\leqslant n-1$, if $c=n$. By the induction hypothesis, MB^k is generated by A and B, hence also M. This completes the proof of the assertion (31).

We are now in a position to prove

THEOREM 2. *Let a, b, c, d be integers such that $ad-bc=1$. Then*

$$\eta\left(\frac{az+b}{cz+d}\right)=\omega\sqrt{cz+d}\cdot\eta(z),\quad \operatorname{Im} z>0, \tag{32}$$

where ω is some 24th root of unity, which depends on a, b, c, d but not on z.

The square-root in (32) is determined by the requirement: $-\frac{1}{2}\pi\leqslant\arg\sqrt{cz+d}\leqslant\frac{1}{2}\pi$.

PROOF. Let M, A, B be respectively the modular transformations

$$z\to\frac{az+b}{cz+d},\qquad z\to-\frac{1}{z},\qquad\text{and}\qquad z\to z+1.$$

The relation $\eta^{24}(Mz)=(cz+d)^{12}\eta^{24}(z)$ holds for $M=A$ by (30), and for $M=B^{\pm 1}$ by (28). If it holds for any modular transformation M, then it holds also for MA and $MB^{\pm 1}$, for

$$\eta^{24}(MAz)=(dz-c)^{12}\eta^{24}(z), \quad MAz=\frac{bz-a}{dz-c}, \tag{33}$$

and

$$\eta^{24}(MB^{\pm 1}z)=(cz\pm c+d)^{12}\eta^{24}(z), \quad MB^{\pm 1}z=\frac{az\pm a+b}{cz\pm c+d}. \tag{34}$$

We have seen that M is a product of powers of A and B; and $A^{-1}=A$. If follows that

$$\eta^{24}(Mz)=(cz+d)^{12}\eta^{24}(z).$$

Taking the 24th root, and noting that $\sqrt{cz+d}$, with the square-root determined as above, is regular for $\operatorname{Im} z>0$, we obtain (32).

Theorem 2 enables us to find $\eta(z)$ in terms of $\eta(z')$, where

$$z'=\frac{az+b}{cz+d}, \quad ad-bc=1. \tag{35}$$

Now every rational point $z=p/q$, $(p,q)=1$, is a singularity of $\eta(z)$. For suppose η were regular at $z=p/q$, $(p,q)=1$. Then there would exist integers r, r', such that $qr'-pr=1$, and therefore a modular transformation $z\to\dfrac{qz-p}{-rz+r'}$, which carries the point $z=\dfrac{p}{q}$ into $z=0$. By (32) there would exist a neighbourhood of the point $z=0$ in which η is regular. By (30) and (27), we would have $\eta(0)=0$, and because of Theorem 2, η also vanishes at an everywhere dense set of rational points on the segment of the real axis which intersects that neighbourhood. Hence, by analytic continuation, η vanishes throughout the half-plane $\operatorname{Im} z>0$, which is impossible.

Lemma. *Given a point* $z=x+iy$ *in the half-plane* $\operatorname{Im} z>0$, *there exists a modular transformation* $z\to z_1=\dfrac{a_1z+b_1}{c_1z+d_1}=x_1+iy_1$, *such that* $y_1\geqslant\dfrac{\sqrt{3}}{2}$, $y_1\geqslant y$.

Proof. As already mentioned, any modular transformation $z\to z'=\dfrac{az+b}{cz+d}$ maps the upper half-plane $\operatorname{Im} z>0$ into itself, since $y'=y|cz+d|^{-2}$, or

$$\frac{1}{y'}=\frac{(cx+d)^2}{y}+c^2y, \quad y>0. \tag{36}$$

Let c, d run through pairs of positive integers, such that $(c,d)=1$. Since the right-hand side of (36) is a positive-definite quadratic form in c, d, it must have a minimum when c, d are restricted to co-prime integral pairs. The minimum may be attained for several pairs of values of c, d. If c_1, d_1 is one such pair, then, since $(c_1, d_1)=1$, there exist integers a_1, b_1, such that $a_1 d_1 - b_1 c_1 = 1$, and the transformation

$$z \to z_1 = \frac{a_1 z + b_1}{c_1 z + d_1} = x_1 + i y_1$$

will obviously transform $y = \operatorname{Im} z$ into its maximum value $y_1 = \operatorname{Im} z_1$.

If we consider the modular transformation

$$z_1 \to z_2 = \frac{a_2 z_1 + b_2}{c_2 z_1 + d_2} = x_2 + i y_2,$$

then, by (36),

$$\frac{1}{y_2} = \frac{(c_2 x_1 + d_2)^2}{y_1} + c_2^2 y_1,$$

and by the maximum property of y_1, we have

$$\frac{1}{y_2} \geqslant \frac{1}{y_1}, \quad \text{for any} \quad c_2, d_2.$$

If we take $c_2 = 1$, and choose the integer d_2, such that $-\frac{1}{2} \leqslant x_1 + d_2 < \frac{1}{2}$, then we get

$$\frac{1}{y_1} \leqslant \frac{1}{y_2} \leqslant \frac{1}{4 y_1} + y_1,$$

or $y_1 \geqslant \dfrac{\sqrt{3}}{2}$, which proves the lemma.

§ 6. The Hardy-Ramanujan formula. The transformation formula for the Dedekind η-function proved in Theorem 2, and the lemma proved thereafter, are of use in the proof of the asymptotic formula for $p(n)$ stated in (1).

THEOREM 3 (G. H. HARDY-S. RAMANUJAN). *If $p(n)$ denotes the number of partitions of n, then*

$$p(n) = \frac{e^{K\lambda_n}}{4\sqrt{3}\,\lambda_n^2} + O\left(\frac{e^{K\lambda_n}}{\lambda_n^3}\right), \quad K = \pi\sqrt{\tfrac{2}{3}}, \quad \lambda_n = \sqrt{n - \tfrac{1}{24}}, \quad n \geqslant 1. \tag{1}$$

PROOF. We have seen in (10) that if

$$f(t)=\sum_{n=0}^{\infty} p(n)\,t^n=\prod_{k=1}^{\infty}(1-t^k)^{-1}, \qquad |t|<1, \tag{9}$$

then

$$p(n)=\frac{1}{2\pi i}\int_C \frac{f(t)}{t^{n+1}}\,dt, \tag{37}$$

where C is a circle with the origin as centre, and radius r, $0<r<1$, in the complex t-plane. If we make a change of variable $t=e^{2\pi iz}$, $z=x+iy$, and use the fact that $f(e^{2\pi iz})=e^{\frac{\pi iz}{12}}\cdot\frac{1}{\eta(z)}$, then we can rewrite (37) as

$$p(n)=\int_L \frac{e^{-2\pi imz}}{\eta(z)}\,dz, \qquad \operatorname{Im} z>0, \tag{38}$$

where $m=n-\frac{1}{24}$, and L a line segment of length 1, parallel to the real axis, from $-\frac{1}{2}+i\varepsilon$ to $\frac{1}{2}+i\varepsilon$, $\varepsilon>0$. We assume C to be such that L is its image under $t\to e^{2\pi iz}$.

We divide L into three parts L_1, L_2, L_3, given by

$$\begin{aligned} &L_1: -\tfrac{1}{2} \leqslant x< -\sqrt{2\varepsilon}, \quad y=\varepsilon,\\ &L_2: -\sqrt{2\varepsilon}\leqslant x< \sqrt{2\varepsilon}, \quad y=\varepsilon,\\ &L_3: \sqrt{2\varepsilon}\leqslant x< \tfrac{1}{2}, \quad y=\varepsilon, \quad \varepsilon^2=(96m)^{-1}, \quad 0<\varepsilon<\tfrac{1}{8}, \end{aligned} \tag{39}$$

where $z=x+iy$, and estimate the integral in (38). We first show that the integral over $L_1\cup L_3$ is $O(e^{\frac{1}{2}K\sqrt{m}})$, $K=\pi\sqrt{\frac{2}{3}}$, $m=n-\frac{1}{24}$.

If $z\in L_1\cup L_3$, there exists a modular transformation

$$z\to z'=\frac{az+b}{cz+d}=x'+iy',$$

such that $y'\geqslant\frac{\sqrt{3}}{2}$, and $y'\geqslant y$, because of the lemma proved in § 5. By (36) we have

$$|cz+d|^2=\frac{y}{y'}\leqslant 1. \tag{40}$$

By Theorem 2, we have

$$\eta(z')=\omega\cdot\sqrt{cz+d}\cdot\eta(z), \qquad \omega^{24}=1.$$

Hence

$$\frac{1}{|\eta(z)|}=\frac{|\sqrt{cz+d}|}{|\eta(z')|}\leqslant\frac{1}{|\eta(z')|}. \tag{41}$$

However, by (9) and (27),

$$\frac{1}{|\eta(z')|} = \left|e^{-\frac{\pi i z'}{12}}\right| \cdot |f(e^{2\pi i z'})|.$$

Since $|e^{2\pi i z'}| = e^{-2\pi y'} \leqslant e^{-\pi\sqrt{3}} < 1$, we have

$$\frac{1}{|\eta(z')|} \leqslant e^{\frac{\pi y'}{12}} \sum_{n=0}^{\infty} p(n)\, e^{-\pi n\sqrt{3}} = O\left(e^{\frac{\pi y'}{12}}\right). \tag{42}$$

From (41) and (42) we obtain

$$\frac{1}{|\eta(z)|} = O\left(e^{\frac{\pi y'}{12}}\right). \tag{43}$$

Now let

$$I = \int\limits_{L_1 \cup L_3} \frac{e^{-2\pi i m z}}{\eta(z)}\, dz. \tag{44}$$

By (36), we have, for $z \in L$,

$$\frac{1}{y'} = \frac{(cx+d)^2}{y} + c^2 y = \frac{(cx+d)^2}{\varepsilon} + c^2 \varepsilon, \tag{45}$$

and if $|c| \geqslant 2$, then $y' \leqslant \dfrac{1}{4\varepsilon}$. Thus it will be sufficient to prove that $|c| \geqslant 2$, in order to show that $y' \leqslant \dfrac{1}{4\varepsilon}$.

We may assume that $c > 0$, for if $c < 0$, we can multiply a, b, c, d by -1, without changing the transformation; and if $c = 0$, then $d = \pm 1$, and (45) gives $\dfrac{1}{y'} = \dfrac{1}{\varepsilon} > 8$, which contradicts the inequality $y' \geqslant \dfrac{\sqrt{3}}{2}$. If $c = 1$, then either $d = 0$ or $d \neq 0$. In the latter case, $|d| \geqslant 1$, since d is an integer, and since $|x| \leqslant \frac{1}{2}$ for $z \in L$, we have $|cx+d| \geqslant |d| - |cx| \geqslant \frac{1}{2}$, so that (45) gives $\dfrac{1}{y'} > \dfrac{1}{4\varepsilon} > 2$, which again contradicts the inequality $y' \geqslant \dfrac{\sqrt{3}}{2}$. If $c = 1$ and $d = 0$, then for $z \in L_1 \cup L_3$, we have $x^2 > 2\varepsilon$, so that (45) again gives $\dfrac{1}{y'} > \dfrac{x^2}{\varepsilon} > 2$, which is a contradiction. Thus $|c| \geqslant 2$, which, as we have observed, implies that $y' \leqslant \dfrac{1}{4\varepsilon}$, where $0 < \varepsilon < \frac{1}{8}$. If we use this bound for y' in (43), we get

$$I = O\left(e^{2\pi m\varepsilon + \frac{\pi}{48\varepsilon}}\right). \tag{46}$$

Since, by choice, $\varepsilon^2=(96m)^{-1}$, where $m=n-\frac{1}{24}$, $n\geqslant 1$, we obtain

$$I=O\left(e^{\frac{K\sqrt{m}}{2}}\right), \qquad K=\pi\sqrt{\tfrac{2}{3}}, \tag{47}$$

where I is defined as in (44).

We now turn to the integral

$$I_0=\int\limits_{L_2}\frac{e^{-2\pi imz}}{\eta(z)}\,dz. \tag{48}$$

Since

$$\frac{1}{\eta(z)}=\frac{e^{-\frac{\pi i}{4}}\sqrt{z}}{\eta\left(-\frac{1}{z}\right)}, \qquad \eta\left(-\frac{1}{z}\right)=\frac{e^{-\frac{\pi i}{12z}}}{f\left(e^{-\frac{2\pi i}{z}}\right)},$$

we have

$$\frac{1}{\eta(z)}=e^{-\frac{\pi i}{4}+\frac{\pi i}{12z}}\cdot\sqrt{z}\cdot f\left(e^{-\frac{2\pi i}{z}}\right).$$

If we put $z=\dfrac{u}{i}$ in (48), we have

$$\begin{aligned} I_0&=-\int\limits_{L_2'}\sqrt{u}\cdot e^{-2\pi mu-\frac{\pi}{12u}}f\left(e^{\frac{2\pi}{u}}\right)du\\ &=-\int\limits_{L_2'}e^{-2\pi g(u)}\cdot\sqrt{u}\cdot f\left(e^{\frac{2\pi}{u}}\right)du, \qquad \sqrt{u}=i \quad \text{for} \quad u=-1. \end{aligned}$$

Here

$$g(u)=mu+\frac{1}{24u},$$

and L_2', the image of L_2 under the mapping $z\to u=iz$, is the line segment which consists of the points $u=\sigma+it$, where $\sigma=-\varepsilon$, $-\sqrt{2\varepsilon}\leqslant t<\sqrt{2\varepsilon}$.

Since

$$f\left(e^{\frac{2\pi}{u}}\right)=1+\sum_{l=1}^{\infty}p(l)\,e^{\frac{2\pi l}{u}}=1+R, \quad \text{say},$$

we have

$$I_0=-\int\limits_{L_2'}e^{-2\pi g(u)}\cdot\sqrt{u}\cdot(1+R)\cdot du.$$

If $u_1=\dfrac{1}{u}=\sigma_1+it_1$, then for $u\in L_2'$, we have

$$\sigma_1=\frac{\sigma}{\sigma^2+t^2}=-\frac{\varepsilon}{\varepsilon^2+t^2}\leqslant-\frac{\varepsilon}{\varepsilon^2+2\varepsilon}=-\frac{1}{\varepsilon+2}<-\frac{1}{3},$$

so that

$$\left|\sum_{l=1}^{\infty} p(l)\, e^{\frac{2\pi m(l)}{u}}\right| \leqslant \sum_{l=1}^{\infty} p(l)\, e^{2\pi m(l)\sigma_1} < \sum_{l=1}^{\infty} p(l)\, e^{-\frac{\pi l}{3}} < \infty,$$

where $m(l)=l-\frac{1}{24}>\frac{1}{2}l$. Therefore

$$e^{-2\pi g(u)}\cdot R = O(e^{2\pi m\varepsilon}) = O\left(e^{\frac{K\sqrt{m}}{4}}\right),$$

since $\sigma=-\varepsilon$ for $u\in L_2'$, and $\varepsilon^2=\dfrac{1}{96m}$, as before. Hence

$$I_0 = -\int_{L_2'} e^{-2\pi g(u)}\cdot\sqrt{u}\cdot du + O\left(e^{\frac{K\sqrt{m}}{4}}\right). \tag{49}$$

From (47) and (49) we get

$$p(n) = -\int_{L_2'} e^{-2\pi g(u)}\cdot\sqrt{u}\cdot du + O\left(e^{\frac{K\sqrt{m}}{2}}\right), \tag{50}$$

where

$$g(u)=mu+\frac{1}{24u}, \qquad m=n-\frac{1}{24}, \qquad K=\pi\sqrt{\frac{2}{3}}, \qquad \varepsilon^2=\frac{1}{96m}.$$

We now seek to evaluate the integral in (50) by applying Cauchy's theorem to a curve $\mathscr{C}$ in the u-plane as indicated in the figure, which is the union of $C_1, L_8, C_0, L_9, C_2, L_7, L_5, -L_2', L_4$ and L_6.

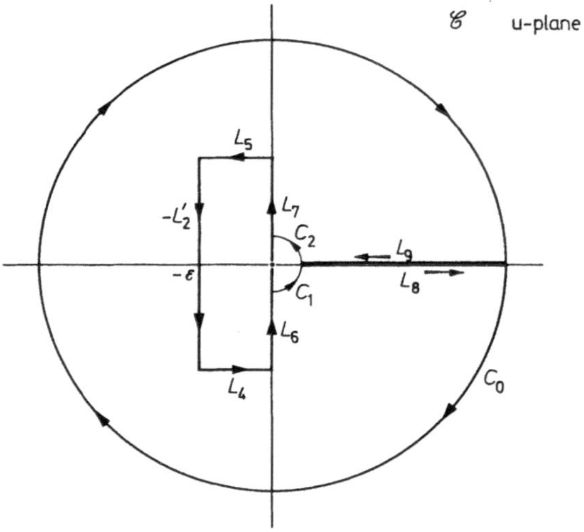

We make a cut in the u-plane along the positive real axis, and note that $\sqrt{u}=i$ for $u=-1$.

The integrals along the (part) circles C_1 and C_2 tend to zero as their radius tends to zero, since $\operatorname{Re} g(u) \geqslant 0$ for $\operatorname{Re} u \geqslant 0$. Now L_6, L_7 as well as L_8, L_9 extend to $u=0$. We denote the extended line segments by $\bar{L}_6, \bar{L}_7, \bar{L}_8, \bar{L}_9$ respectively.

If $u \in \bar{L}_6 \cup \bar{L}_7$, then $\operatorname{Re} u=0$, and $\operatorname{Re} g(u)=0$, so that $|e^{-2\pi g(u)}|=1$. Hence

$$\Big| - \int_{\bar{L}_6 \cup \bar{L}_7} e^{-2\pi g(u)} \cdot \sqrt{u} \cdot du \Big| = O(1). \tag{51}$$

If $u \in L_4 \cup L_5$, then $u=\sigma \pm i\sqrt{2\varepsilon}$, $-\varepsilon<\sigma<0$, and

$$\sigma_1 = \operatorname{Re} \frac{1}{u} = \frac{\sigma}{\sigma^2+t^2} > -\frac{\varepsilon}{\sigma^2+2\varepsilon} > -\frac{1}{2},$$

so that $\operatorname{Re} g(u) > -m\varepsilon - \frac{1}{48}$, and

$$|e^{-2\pi g(u)}| = O(e^{2\pi m\varepsilon}),$$

which implies that

$$\Big| - \int_{L_4 \cup L_5} e^{-2\pi g(u)} \cdot \sqrt{u} \cdot du \Big| = O(e^{2\pi m \varepsilon}) = O\left(e^{\frac{K\sqrt{m}}{4}}\right). \tag{52}$$

It remains for us to consider the integral

$$-\int_{\bar{L}_8 \cup C_0 \cup \bar{L}_9} e^{-2\pi g(u)} \cdot \sqrt{u} \cdot du,$$

where $\sqrt{u}$ is taken positive along $\bar{L}_9$ and negative along $\bar{L}_8$. If we set

$$\alpha = 2\pi\left(n - \frac{1}{24}\right), \quad \beta = \frac{\pi}{12}, \quad \mathscr{K} = \bar{L}_8 \cup C_0 \cup \bar{L}_9,$$

then we have to evaluate the integral

$$-\int_{\mathscr{K}} e^{-\alpha u - \beta/u} \cdot \sqrt{u} \cdot du. \tag{53}$$

This is, however, the derivative, with respect to α of the integral

$$J = \int_{\mathscr{K}} e^{-\alpha u - \beta/u} \frac{du}{\sqrt{u}}. \tag{54}$$

If we put $v=\sqrt{\alpha u}+\sqrt{\beta/u}$, $w=-\sqrt{\alpha u}+\sqrt{\beta/u}$, then $2\sqrt{u}=\dfrac{v-w}{\sqrt{\alpha}}$, so that $\dfrac{1}{\sqrt{u}} = \dfrac{1}{\sqrt{\alpha}}\left(\dfrac{dv}{du} - \dfrac{dw}{du}\right)$. If we choose the radius of the circle C_0 to be $\sqrt{\beta/\alpha}$, then

$$J = \int_{\mathscr{K}} e^{-\alpha u - \beta/u} \cdot \frac{1}{\sqrt{\alpha}} \left(\frac{dv}{du} - \frac{dw}{du} \right) \cdot du$$

$$= \frac{e^{2\sqrt{\alpha\beta}}}{\sqrt{\alpha}} \int_{\mathscr{K}_1} e^{-v^2} dv - \frac{e^{-2\sqrt{\alpha\beta}}}{\sqrt{\alpha}} \int_{\mathscr{K}_2} e^{-w^2} dw = J_1 - J_2, \text{ say,} \tag{55}$$

where $\mathscr{K}_1$ is the image of $\mathscr{K}$ under the map $u \to v$, and $\mathscr{K}_2$ the image of $\mathscr{K}$ under the map $u \to w$.

Clearly

$$J_1 = \frac{e^{2\sqrt{\alpha\beta}}}{\sqrt{\alpha}} \int_{-\infty}^{\infty} e^{-v^2} dv = \sqrt{\frac{\pi}{\alpha}}\, e^{2\sqrt{\alpha\beta}},$$

and

$$J_2 = \frac{e^{-2\sqrt{\alpha\beta}}}{\sqrt{\alpha}} \int_{-\infty}^{\infty} e^{-w^2} dw = \sqrt{\frac{\pi}{\alpha}}\, e^{-2\sqrt{\alpha\beta}},$$

so that

$$J = J_1 - J_2 = \sqrt{\frac{\pi}{\alpha}} \left(e^{2\sqrt{\alpha\beta}} - e^{-2\sqrt{\alpha\beta}} \right) = 2\sqrt{\frac{\pi}{\alpha}} \sinh(2\sqrt{\alpha\beta}),$$

and, by (54)–(50), we get

$$p(n) = \frac{d}{d\alpha} \left(\sqrt{\frac{\pi}{\alpha}} \left(e^{2\sqrt{\alpha\beta}} - e^{-2\sqrt{\alpha\beta}} \right) \right) + O\left(e^{\frac{K\sqrt{m}}{2}} \right), \tag{56}$$

where

$$\alpha = 2\pi\left(n - \frac{1}{24}\right), \quad \beta = \frac{\pi}{12}, \quad 2\sqrt{\alpha\beta} = \pi\sqrt{\frac{2}{3}} \cdot \sqrt{n - \frac{1}{24}} = K\sqrt{n - \frac{1}{24}}.$$

If we carry out the differentiation in (56), take the leading term, and replace $\sqrt{n - \frac{1}{24}}$ by λ_n, we get the formula

$$p(n) = \frac{e^{K\lambda_n}}{4\sqrt{3}\,\lambda_n^2} + O\left(\frac{e^{K\lambda_n}}{\lambda_n^3} \right),$$

as claimed in Theorem 3.

§ 7. Rademacher's identity. The proof of Theorem 3 was carried out in three steps. We first showed that the main contribution to the integral for $p(n)$, given in (38), comes from that part of the contour L which lies in a neighbourhood of $z=0$, namely L_2. We next showed that in that neighbourhood one can replace $\eta(-1/z)$ by $e^{-\frac{\pi i}{12z}}$ with a sufficiently small error. We then evaluated the integral resulting from such a replacement. In order to obtain a convergent series for $p(n)$, we have to carry out this procedure, not only for the singularity $z=0$ of $f(e^{2\pi iz})$, given in (9), but for every point $z=\frac{p}{q}$, $(p,q)=1$, on a segment of the real axis of length 1.

Let k be an arbitrary positive integer, $k>1$. We now choose L to be the line segment

$$-\frac{1}{2k}\leqslant x<1-\frac{1}{2k},\quad y=\frac{1}{k^2},\quad z=x+iy.$$

We shall ultimately let $k\to\infty$, and all the O's in the sequel will refer to k. We keep n as a fixed, positive integer.

Let p and q be integers, such that $(p,q)=1$, and $0<q\leqslant k$. Consider the fractions $\frac{p}{q}$, such that $0\leqslant\frac{p}{q}\leqslant 1$. They define the *Farey sequence* of order k. We consider special neighbourhoods of the points $z=\frac{p}{q}$.

Let $\alpha_{p,q}$ be an interval of the real axis, in the z-plane, such that $x\in\alpha_{p,q}$ if and only if

$$\left|x-\frac{p}{q}\right|\leqslant\frac{1}{2qk}. \tag{57}$$

Thus $\alpha_{0,1}$ is the interval $\left[-\frac{1}{2k},\frac{1}{2k}\right]$, and $\alpha_{1,1}$ is the interval $\left[1-\frac{1}{2k},1+\frac{1}{2k}\right]$. *These intervals* $(\alpha_{p,q})$ *do not overlap.* That is, they have no interior point in common. They may have an end-point in common. For suppose, if possible, that $\alpha_{p,q}$ and $\alpha_{p',q'}$, where $\frac{p}{q}\neq\frac{p'}{q'}$, overlap. Then there exists a point x which is an interior point of both the intervals. Hence

$$|u|<\frac{1}{2k},\qquad |u'|<\frac{1}{2k},$$

where $u=qx-p$, $u'=q'x-p'$. Now

$$|qp'-pq'|=|qu'-uq'|<(q+q')\frac{1}{2k}\leqslant 1,$$

so that $|qp'-pq'|<1$, which implies that $qp'-pq'=0$, since it must be an integer. Hence $\frac{p}{q}=\frac{p'}{q'}$, which contradicts the assumption.

We next observe that if $k>1$, the segment of the real axis, $-\frac{1}{2k}\leqslant x<1-\frac{1}{2k}$, contains all the intervals $\alpha_{p,q}$ such that $0\leqslant\frac{p}{q}<1$. For the highest permissible value of p/q is $(k-1)/k$, and $\{(k-1)/k\}-1/(2k^2)<1-1/(2k)$, if $k>1$. And the intervals $(\alpha_{p,q})$ cannot overlap.

We now split up the line of integration L, which consists of the points $z=x+iy$, with $-\frac{1}{2k}\leqslant x<1-\frac{1}{2k}$ and $y=\frac{1}{k^2}$, $k>1$, into two sets, namely $\bigcup_{p,q}\beta_{p,q}$, $0\leqslant\frac{p}{q}<1$, and β_0, where $\beta_{p,q}$ is an interval in L which consists of the points $z=x+iy$, $x\in\alpha_{p,q}$, $y=\frac{1}{k^2}$, while β_0 consists of the points $z=x+iy$, $z\in L$, $x\notin\alpha_{p,q}$ for $0\leqslant\frac{p}{q}<1$, and $y=\frac{1}{k^2}$.

We shall first show that

$$\int_{\beta_0}\frac{e^{-2\pi imz}}{\eta(z)}\,dz\to 0,\quad \text{as}\quad k\to\infty. \tag{58}$$

For any point z, with $\operatorname{Im}z>0$, there exists, by the lemma proved in § 5, a point z', such that

$$z'=x'+iy'=\frac{az+b}{cz+d},\qquad ad-bc=1,$$

and $y'\geqslant\frac{\sqrt{3}}{2}$. Further $y'=y|cz+d|^{-2}$, and by Theorem 2, $\eta(z')=\omega\sqrt{cz+d}\cdot\eta(z)$, while $\eta(z)=\frac{e^{\pi iz/12}}{f(e^{2\pi iz})}$, where $f(z)=\sum_{n=0}^{\infty}p(n)z^n$, $|z|<1$. Thus

$$\frac{1}{|\eta(z)|}=\frac{|cz+d|^{\frac{1}{2}}}{|\eta(z')|}\leqslant\left(\frac{y}{y'}\right)^{\frac{1}{4}}e^{\pi y'/12}\sum_{j=0}^{\infty}p(j)e^{-2\pi jy'},$$

as in (40) and (42). Since $y'\geqslant\frac{\sqrt{3}}{2}$, we have $\sum_{j=0}^{\infty}p(j)e^{-2\pi jy'}=O(1)$. Further

$$\frac{1}{y'}=\frac{(cx+d)^2}{y}+c^2y=k^2(cx+d)^2+\frac{c^2}{k^2},\quad \text{for}\quad z\in L.$$

We shall see that

$$y' < 4.$$

For if

$$\frac{1}{y'} = k^2(cx+d)^2 + \frac{c^2}{k^2} \leqslant \frac{1}{4},$$

then $0 \leqslant c \leqslant \frac{1}{2}k$, and $|cx+d| \leqslant \frac{1}{2k}$. If $c=0$, then $1 \leqslant d \leqslant \frac{1}{2k}$, which is impossible (note the remark preceding (31)); so that $1 \leqslant c \leqslant \frac{1}{2}k$, $\left| x + \frac{d}{c} \right| < \frac{1}{2kc}$. Since $x = \operatorname{Re} z$, $z \in \beta_0$, it follows that either $-\frac{d}{c} < 0$, or $-\frac{d}{c} \geqslant 1$. That is, $x < \frac{1}{2k}$ or $x > 1 - \frac{1}{2k}$, which contradicts the fact that $\frac{1}{2k} \leqslant x \leqslant 1 - \frac{1}{2k}$ since $x \in \beta_0$. Hence $y' < 4$, and it follows that

$$\frac{1}{|\eta(z)|} = O(y^{\frac{1}{4}}) = O(k^{-\frac{1}{2}}).$$

Since the measure of the set β_0 is less than 1, and $e^{2\pi m y} = O(1)$, for fixed n, as $y \to 0+$, we have

$$\int_{\beta_0} \frac{e^{-2\pi i m z}}{\eta(z)} dz = O(k^{-\frac{1}{2}}) = o(1),$$

as $k \to \infty$, which proves (58).

We shall next consider

$$\int_{\beta_{p,q}} \frac{e^{-2\pi i m z}}{\eta(z)} dz, \qquad 0 \leqslant \frac{p}{q} < 1, \qquad (p,q) = 1\,.$$

Since $(p,q)=1$, there exist an infinity of pairs of integers a and b, such that $-ap-bq=1$. Take that pair (a,b) for which a, greater than zero, has its minimum value. Let $c=q$, $d=-p$, so that

$$z' = \frac{az+b}{cz+d} = \frac{az+b}{qz-p} = x' + iy'.$$

We cannot now assert that y' is necessarily $\geqslant \frac{1}{2}\sqrt{3}$. We have

$$\frac{1}{\eta(z)} = \omega_{p,q} \cdot \sqrt{cz+d} \cdot \frac{1}{\eta(z')} = \omega_{p,q} \cdot \sqrt{cz+d} \cdot e^{-\frac{\pi i z'}{12}} \sum_{j=0}^{\infty} p(j) e^{2\pi i j z'}$$

$$= \omega_{p,q} \cdot \sqrt{cz+d} \cdot e^{-\frac{\pi i z'}{12}} (1 + R_{p,q}), \quad \text{say.} \tag{59}$$

We shall show that

$$\sum_{p,q} \int_{\beta_{p,q}} |e^{-2\pi imz}| \left| \omega_{p,q} \cdot \sqrt{cz+d} \cdot e^{-\frac{\pi iz'}{12}} R_{p,q} \right| \cdot |dz| = O(k^{-\frac{1}{2}}), \qquad (60)$$

as $k \to \infty$.

Since $y' = y|cz+d|^{-2}$, we have

$$\left| \omega_{p,q} \cdot \sqrt{cz+d} \cdot e^{-\frac{\pi iz'}{12}} \cdot R_{p,q} \right| \leqslant \left(\frac{y}{y'} \right)^{\frac{1}{4}} \sum_{j=1}^{\infty} p(j) e^{-2\pi\left(j-\frac{1}{24}\right)y'}. \qquad (61)$$

We shall show that y' has a positive lower bound for any p, q. For $\frac{1}{y'} = k^2(cx+d)^2 + \frac{c^2}{k^2}$, and $c = q$, $d = -p$ by assumption, while $z = x + iy \in \beta_{p,q}$ implies that

$$\left| x - \frac{p}{q} \right| \leqslant \frac{1}{2qk}, \quad \text{or} \quad |qx - p| = |cx + d| \leqslant \frac{1}{2k}.$$

Hence $\frac{1}{y'} \leqslant \frac{1}{4} + \frac{q^2}{k^2} \leqslant \frac{5}{4}$, since $0 < q \leqslant k$. Thus $y' \geqslant \frac{4}{5}$, so that $\left(\frac{y}{y'} \right)^{\frac{1}{4}} = O(y^{\frac{1}{4}}) = O(k^{-\frac{1}{2}})$. Hence, by (61),

$$\int_{\beta_{p,q}} |e^{-2\pi imz}| \cdot \left| \omega_{p,q} \cdot \sqrt{cz+d} \cdot e^{-\frac{\pi iz'}{12}} R_{p,q} \right| \cdot |dz| = |\beta_{p,q}| \cdot O(k^{-\frac{1}{2}}), \quad (62)$$

as $k \to \infty$, uniformly in p and q, for fixed n. Here $|\beta_{p,q}|$ is the measure of the interval $\beta_{p,q}$.

Since the intervals $(\beta_{p,q})$ do not overlap, and since their total measure is at most one, we infer that

$$\sum_{p,q} \int_{\beta_{p,q}} |e^{-2\pi imz}| \cdot \left| \omega_{p,q} \cdot \sqrt{cz+d} \cdot e^{-\frac{\pi iz'}{12}} R_{p,q} \right| \cdot |dz| = O(k^{-\frac{1}{2}}),$$

as $k \to \infty$, which proves (60).

In view of (59) it remains for us to evaluate the integral in the formula

$$p(n) = \sum_{p,q} \omega_{p,q} \int_{\beta_{p,q}} e^{-2\pi imz - \frac{\pi iz'}{12}} \cdot \sqrt{cz+d} \cdot dz + O(k^{-\frac{1}{2}}). \qquad (63)$$

Now

$$z' = \frac{az+b}{cz+d} = \frac{c(az+b)}{c(cz+d)} = \frac{a(cz+d)-1}{c(cz+d)} = \frac{a}{c} - \frac{1}{c(cz+d)},$$

so that

$$-2\pi i m z - \frac{\pi i z'}{12} = -2\pi i m z - \frac{\pi i a}{12c} + \frac{\pi i}{12c(cz+d)}. \tag{64}$$

Set

$$cz+d = qz-p = -iw, \tag{65}$$

so that

$$z = \frac{1}{q}(p-iw), \qquad z' = \frac{a}{c} - \frac{i}{cw} = \frac{a}{q} - \frac{i}{qw}. \tag{66}$$

Then

$$p(n) = -\sum_{p,q} \rho_{p,q} \int_{\mathscr{L}} e^{-2\pi i m \frac{p}{q} - 2\pi g(w)} \sqrt{w}\, \frac{dw}{q} + O(k^{-\frac{1}{2}}), \tag{67}$$

where

$$\rho_{p,q} = \omega_{p,q} e^{\frac{\pi i}{4} - \frac{\pi i a}{12q}}, \qquad |\rho_{p,q}| = 1, \tag{68}$$

$$g(w) = \frac{m}{q} w + \frac{1}{24qw}, \; \sqrt{w} = i \quad \text{for} \quad w = -1. \tag{69}$$

Since $\beta_{p,q}$ consists of points $z = x+iy$, such that

$$\frac{p}{q} - \frac{1}{2qk} \leqslant x \leqslant \frac{p}{q} + \frac{1}{2qk}, \qquad y = \frac{1}{k^2},$$

the path of integration $\mathscr{L} = \mathscr{L}(p,q)$ in (67) is the line segment which consists of the points w, with $\operatorname{Re} w = -\frac{q}{k^2}$ and $-\frac{1}{2k} \leqslant \operatorname{Im} w \leqslant \frac{1}{2k}$. As in the evaluation of the integral in (50), we extend the path of integration to a contour consisting of a circle $\mathscr{C}$, with the origin as centre, and with a cut along the positive real axis in the w-plane, as shown in the figure next page.

We shall show that

$$-\sum_{p,q} \rho_{p,q} \int_{\mathscr{L}' \cup \mathscr{L}''} e^{-2\pi i m \frac{p}{q} - 2\pi g(w)} \sqrt{w} \cdot \frac{dw}{q} = O(k^{-\frac{1}{2}}). \tag{70}$$

To prove this, we first note that for $w \in \mathscr{L}' \cup \mathscr{L}''$, $|w| = O(k^{-1})$. Secondly, if $w \in \mathscr{L}'$, then $w = \sigma \pm \frac{i}{2k}$, $-\frac{q}{k^2} < \sigma < 0$, so that

$$\operatorname{Re} \frac{1}{w} = \frac{\sigma}{\sigma^2 + \frac{1}{4k^2}} > -\frac{q}{k^2} \cdot \frac{1}{\sigma^2 + \frac{1}{4k^2}} \geqslant -4q, \quad \text{or} \quad \operatorname{Re} \frac{1}{24qw} > -\frac{1}{6}.$$

By (69) it follows that $|e^{-2\pi g(w)}| = O(1)$, since m is fixed. If, on the other hand, $w \in \mathscr{L}''$, then w is imaginary, so is $g(w)$ and $|e^{-2\pi g(w)}| = O(1)$.

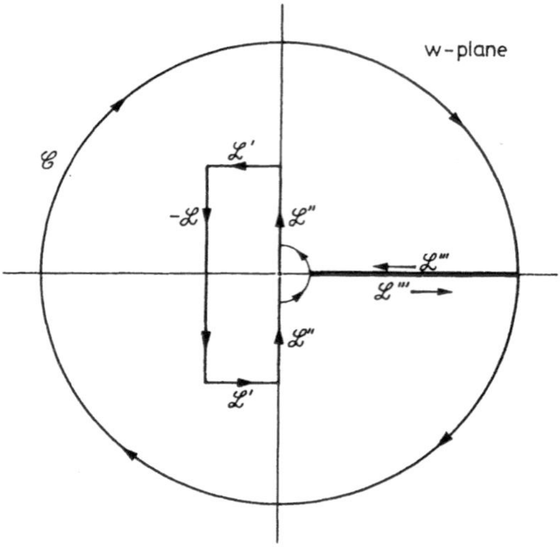

Hence the integrand in (70) is $O(k^{-\frac{1}{2}}q^{-1})$, while the measure of $\mathscr{L}' \cup \mathscr{L}''$ is $O(k^{-1})$. Therefore the integral in (70) is $O(k^{-\frac{3}{2}}q^{-1})$. The number of such integrals in the sum (70) equals the number of fractions $\frac{p}{q}$, where $(p,q)=1$, and $0 \leqslant \frac{p}{q} < 1$. For any q, there are at most q such fractions. Hence, for a fixed q,

$$-\sum_{p} \rho_{p,q} \int_{\mathscr{L}' \cup \mathscr{L}''} e^{-2\pi i m \frac{p}{q} - 2\pi g(w)} \sqrt{w}\, \frac{dw}{q} = q \cdot O(k^{-\frac{3}{2}} q^{-1}) = O(k^{-\frac{3}{2}}).$$

If we now sum over q, $0 < q \leqslant k$, we obtain (70).

In order to calculate the sum of integrals

$$-\sum_{p,q} \rho_{p,q} \int_{\mathscr{C} \cup \mathscr{L}'''} e^{-2\pi i m \frac{p}{q} - 2\pi g(w)} \sqrt{w}\, \frac{dw}{q},$$

we proceed as we did with the evaluation of (53). If we now write

$$\alpha = \frac{2\pi}{q}\left(n - \frac{1}{24}\right) = \frac{2\pi m}{q}, \qquad \beta = \frac{\pi}{12q}, \qquad \mathscr{C}_1 = \mathscr{C} \cup \mathscr{L}''',$$

then, by (67) and (70), we have

$$p(n) = O(k^{-\frac{1}{2}}) - \sum_{p,q} \rho_{p,q}\, e^{-2\pi i m \frac{p}{q}} \int_{\mathscr{C}_1} e^{-\alpha w - \beta/w} \sqrt{w}\, \frac{dw}{q}. \tag{71}$$

Since the integral here does *not* depend on p, we set

$$A_q(n) = \sum_p \rho_{p,q} e^{-2\pi i m \frac{p}{q}}, \tag{72}$$

where

$$0 \leqslant \frac{p}{q} < 1, (p,q) = 1, m = n - \frac{1}{24}, \rho_{p,q} = \omega_{p,q} e^{\frac{\pi i}{4} - \frac{\pi i a}{12 q}}, (\omega_{p,q})^{24} = 1,$$

and write (71) as

$$p(n) = O(k^{-\frac{1}{2}}) - \sum_{q=1}^{k} A_q(n) \int_{\mathscr{C}_1} e^{-\alpha w - \beta/w} \cdot \sqrt{w} \cdot \frac{dw}{q}, k > 1.$$

Evaluating the integral as in (54), we get

$$p(n) = O(k^{-\frac{1}{2}}) + \sum_{q=1}^{k} \frac{A_q(n)}{q} \cdot \frac{d}{d\alpha} \left(\sqrt{\frac{\pi}{\alpha}} (e^{2\sqrt{\alpha\beta}} - e^{-2\sqrt{\alpha\beta}}) \right). \tag{73}$$

If we now let $k \to \infty$, we get the convergent series

$$p(n) = \sum_{q=1}^{\infty} \frac{A_q(n)}{q} \frac{d}{d\alpha} \left(\sqrt{\frac{\pi}{\alpha}} \cdot 2 \sinh(2\sqrt{\alpha\beta}) \right), \tag{74}$$

from which we infer

THEOREM 4 (RADEMACHER). *If $p(n)$ denotes the number of partitions of the positive integer n, then*

$$p(n) = \frac{1}{\pi\sqrt{2}} \sum_{q=1}^{\infty} A_q(n) \cdot q^{\frac{1}{2}} \cdot \frac{d}{dn} \frac{\sinh\left(\frac{K}{q} \sqrt{n - \frac{1}{24}} \right)}{\sqrt{n - \frac{1}{24}}}, \tag{75}$$

where $K = \pi\sqrt{\frac{2}{3}}$, and $A_q(n)$ is defined as in (72).

Notes on Chapter VII

The presentation here follows C.L. Siegel's *Lectures on analytic number theory*, New York University (1945), Ch. III. It incorporates several improvements made by him in subsequent lectures. A fascinating exposition of the asymptotic theory of partitions is given by G. H. Hardy in *Ramanujan*, Ch. VIII. A leisurely and interesting account of the theory is given by H. Rademacher in his *Lectures on analytic number theory*, Tata Institute of Fundamental Research, Bombay (1955).

§§ 1—4. The elementary theory of partitions goes back to Euler. See Dickson's *History*, II, 103. For an exposition one may refer to P. A. MacMahon, *Combinatory Analysis* (Cambridge, 1916), II; Hardy and Wright, *Theory of numbers*, 2nd edition, Ch.XIX; Sylvester's *Collected papers* (Cambridge, 1908), II, 119—175.

For the asymptotic theory, the original sources are G. H. Hardy and S. Ramanujan, *Proc. London Math. Soc.* (2) 17 (1918), 75—115 (No. 36 of Ramanujan's *Collected papers*); J.V. Uspensky, *Bull. de l'Acad. des Sciences de l'URSS*, (6) 14 (1920), 199—218; H. Rademacher, *Proc. London Math. Soc.* (2) 43 (1937), 241—254; *Proc. Nat. Acad. Sci. USA* 23 (1937), 78—84; *Annals of Math.* (2) 44 (1943), 416—422.

The Hardy-Ramanujan paper contains "the circle method" which has had many applications.

The following comment by Hardy on Uspensky's paper may be noted: "Uspensky's paper was published a little after ours, and we developed the solution much further, so that his proof of

$$p(n) = \frac{1}{2\pi\sqrt{2}} \frac{d}{dn}\left(\frac{e^{K\lambda_n}}{\lambda_n}\right) + O\left(e^{Hn^{\frac{1}{2}}}\right), \quad H < K,$$

which is simpler than ours, has been noticed less than it deserves." The complete formula proved by Hardy and Ramanujan (formula (1.74) of their paper, loc. cit., 285), however, closely resembles Rademacher's formula (2), but contains an error term $O(n^{-\frac{1}{4}})$.

§ 5. Several proofs of (30) are known. See C. L. Siegel, *Mathematika* 1 (1954), 4; *Gesammelte Abhandlungen*, III, 188; *Lectures on advanced analytic number theory*, Tata Institute of Fundamental Research, Bombay (1961), Ch. 1. Siegel's argument has been extended by Rademacher, *J. Indian Math. Soc.* 19 (1955), 25—30.

For a succinct account of theta-functions, which includes the Dedekind η-function, see C. L. Siegel's lecture notes on *Analytische Zahlentheorie*, Göttingen (1963—64), II, 1—63.

For the explicit determination of ω in Theorem 2, in the manner of Hermite's determination of the eighth root of unity, which occurs in the transformation formula for the theta-functions, the original sources (as communicated to the author by Professor Siegel) are: Th. Molien, *Berichte über die Verhandlungen der Königlich Sächsischen Gesellschaft der Wissenschaften zu Leipzig, Mathematisch-Physische Classe*, 37 (1885), 25—38; L. Kiepert, *Math. Annalen*, 26 (1886), 369—454, particularly Abschnitt III, 404—422. As further references, see F. Klein, *Math. Annalen*, 26 (1886), 455—464; F. Klein and R. Fricke, *Vorlesungen über die Theorie der Elliptischen Modulfunktionen*, (1892), II, 70, 81; H. Weber, *Lehrbuch der Algebra*, (1908), III, § 38, and the references given there.

§§ 6—7. This formulation of the proof of Theorem 3 and of Theorem 4 is due to C. L. Siegel, particularly the evaluation of the integrals. As he remarks, formula (1) may be written more simply as

$$p(n) = \frac{e^{K\sqrt{n}}}{4\sqrt{3}n}\left(1 + O\left(\frac{1}{\sqrt{n}}\right)\right).$$

For an account of Farey sequences, see the author's *Introduction*, Ch. I.

It is possible to estimate the error involved in calculating $p(n)$ by taking N terms of the convergent series in (73). For this purpose the crude estimate $|A_q(n)| \leqslant q$ was improved by D. H. Lehmer to $|A_q(n)| < 2q^{\frac{5}{6}}$, *Trans. Amer. Math. Soc.* 43 (1938), 292. Lehmer also showed that $A_q(n)$ is multiplicative. For remarks about the numerical calculations, see Hardy's *Ramanujan*, 130—131.

Rademacher, in his Bombay Lectures, has described the work of Atle Selberg on $A_q(n)$. It follows from the work of Selberg that $A_q(n) = O(q^{\frac{1}{2}+\varepsilon})$, $\varepsilon > 0$.

Various expressions for $\rho_{p,q}$ have been given. See Rademacher, *J. London Math. Soc.* 7 (1932), 14—19.

For the congruence properties of the partition function, see Ramanujan's *Collected papers*, loc. cit., Nos. 25, 28, and 30. See also Hardy's *Ramanujan*, Ch. VI.

Chapter VIII

Dirichlet's divisor problem

§ 1. The average order of the divisor function. Let $d(n)$ denote the number of positive divisors of the positive integer n. Let

$$E(x) = \sum_{n \leqslant x} d(n) - x \log x - (2\gamma - 1)x, \quad x \geqslant 1, \tag{1}$$

where γ is Euler's constant. It is known, after Dirichlet, that

$$E(x) = O(x^{\frac{1}{2}}), \quad \text{as} \quad x \to \infty. \tag{2}$$

Dirichlet's divisor problem is to determine the precise order of magnitude of $E(x)$ as $x \to \infty$. Let θ denote the smallest value of ξ, such that $E(x) = O(x^{\xi + \varepsilon})$, for every $\varepsilon > 0$. G. Voronoi was the first to prove that $\theta \leqslant \frac{1}{3}$. On the other hand, Hardy and Landau proved independently that $\theta \geqslant \frac{1}{4}$. The conjecture that $\theta = \frac{1}{4}$ has not yet been proved or disproved.

We shall prove in this chapter Voronoi's theorem that

$$E(x) = O(x^{\frac{1}{3}} \log x), \quad \text{as} \quad x \to \infty. \tag{3}$$

We shall also prove Hardy's result that

$$\limsup_{x \to \infty} \frac{E(x)}{x^{\frac{1}{4}}} = +\infty, \tag{4}$$

as well as Ingham's result that

$$\liminf_{x \to \infty} \frac{E(x)}{x^{\frac{1}{4}}} = -\infty. \tag{5}$$

The proof depends on a study of the asymptotic behaviour of a sum of the form

$$D^{\rho}(x) = \frac{1}{\Gamma(\rho + 1)} \sum_{n \leqslant x} d(n)(x - n)^{\rho}, \qquad \rho \geqslant 0, \tag{6}$$

as $x \to \infty$, and particularly on the fact that such a sum has an expansion in a series of special functions. By an application of a theorem of

A. Zygmund on the *equiconvergence* of such series with trigonometric series, we shall deduce the identity

$$\sum_{n\leqslant x} d(n)=\tfrac{1}{4}+x\log x+(2\gamma-1)x-\sum_{n=1}^{\infty} d(n)\left(\frac{x}{n}\right)^{\frac{1}{2}} F_1\{4\pi(nx)^{\frac{1}{2}}\}, \tag{7}$$

for $x>1$, x non-integral, and $F_1(x)=Y_1(x)+\dfrac{2}{\pi}K_1(x)$, where $Y_1(x)$ and $K_1(x)$ stand for the well-known Bessel functions. The method of proof followed here is the one developed by the author and Raghavan Narasimhan to prove similar results for a wide class of arithmetical functions.

§ 2. An application of Perron's formula. The Dirichlet series $\sum_{n=1}^{\infty} d(n)n^{-s}$, $s=\sigma+it$, converges absolutely for $\sigma>1$, and the sum-function is $\zeta^2(s)$, the square of Riemann's zeta-function. It satisfies the functional equation

$$\pi^{-s}\Gamma^2(\tfrac{1}{2}s)\zeta^2(s)=\pi^{s-1}\Gamma^2(\tfrac{1}{2}-\tfrac{1}{2}s)\zeta^2(1-s), \tag{8}$$

which is obtained by squaring the functional equation of $\zeta(s)$.

By a general form of Perron's formula for the partial sum of the coefficients of a Dirichlet series, we have

$$D^{\rho}(x)=\frac{1}{2\pi i}\int_{\beta-i\infty}^{\beta+i\infty}\frac{\Gamma(s)\zeta^2(s)x^{s+\rho}}{\Gamma(\rho+1+s)}\,ds, \tag{9}$$

where $\rho>0$, $\beta\geqslant 1+\delta>1$, $x>0$. We shall assume that β is not an integer. The integrand is a meromorphic function whose poles are contained among the integers $\leqslant 1$.

Let R be a fixed number greater than zero; and let r be positive, non-integral, and sufficiently large; in particular, let $r>\rho+1$. Having chosen the r, we keep it fixed.

We shall move the line of integration in (9) from $s=\beta+it$, $-\infty<t<+\infty$, to the broken line L through $1-\beta-i\infty$, $1-\beta-iR$, $1-r-iR$, $1-r+iR$, $1-\beta+iR$, and $1-\beta+i\infty$, in that order. This we shall do by applying Cauchy's theorem of residues to the simple curve with corners at $\beta-iT$, $1-\beta-iT$, $1-\beta-iR$, $1-r-iR$, $1-r+iR$, $1-\beta+iR$, $1-\beta+iT$ and $\beta+iT$, in that order, where $T>R$, and then letting $T\to\infty$.

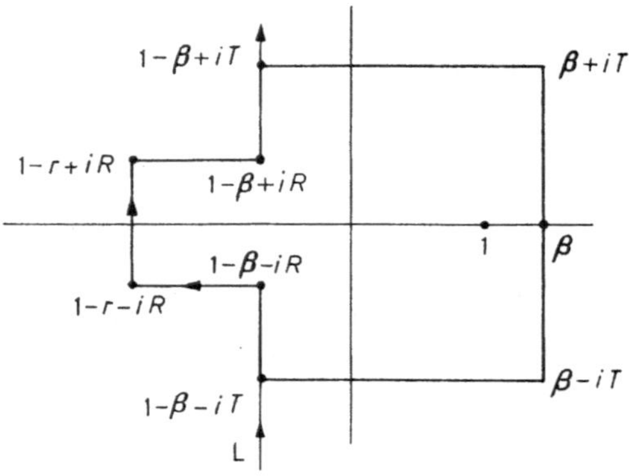

It is easy to see that the contribution to the integral from the two sides of the figure, namely $\sigma \pm iT$, $1-\beta<\sigma<\beta$, tends to zero as $T\to\infty$. For on the line $s=\beta+it$, we have $\zeta^2(s)=O(1)$; and since

$$|\Gamma(x+iy)| \sim e^{-\frac{1}{2}\pi|y|} \cdot |y|^{x-\frac{1}{2}} \cdot \sqrt{(2\pi)}, \tag{10}$$

as $|y|\to\infty$, uniformly in $-\infty<x_1\leqslant x\leqslant x_2<+\infty$, we have

$$\frac{\Gamma(s)}{\Gamma(\rho+1+s)}=O\left(|t|^{-\rho-1}\right)=o(1), \quad \text{if} \quad \rho>-1, \tag{11}$$

as $|t|\to\infty$. Hence for $s=\beta+it$, $-\infty<t<+\infty$,

$$\frac{\Gamma(s)\zeta^2(s)x^{s+\rho}}{\Gamma(\rho+1+s)}=o(1), \quad \text{as} \quad |t|\to\infty. \tag{12}$$

This estimate holds also on the line $s=1-\beta+it$, $-\infty<t<+\infty$, if we use equation (8), and observe that

$$\left|\frac{\Gamma(1-\beta+it)\zeta^2(1-\beta+it)x^{1-\beta+\rho+it}}{\Gamma(\rho+2-\beta+it)}\right|$$
$$=\left|\frac{\Gamma(1-\beta-it)\zeta^2(1-\beta-it)x^{1-\beta+\rho-it}}{\Gamma(\rho+2-\beta-it)}\right|$$
$$=\left|\frac{\Gamma(1-\beta-it)\zeta^2(\beta+it)\Gamma^2(\frac{1}{2}\beta+\frac{1}{2}it)\pi^{-\beta-it}x^{1-\beta+\rho-it}}{\Gamma(\rho+2-\beta-it)\Gamma^2(\frac{1}{2}-\frac{1}{2}\beta-\frac{1}{2}it)\pi^{\beta-1+it}}\right|$$
$$=O\left(|t|^{2\beta-\rho-2}\right)=o(1), \quad \text{as} \quad |t|\to\infty, \quad \text{if} \quad \rho>2\beta-2.$$

We can now infer, by the Phragmén-Lindelöf principle, that

$$\frac{\Gamma(s)\zeta^2(s)x^{s+\rho}}{\Gamma(\rho+1+s)} = o(1), \quad \text{as} \quad |t|\to\infty, \tag{13}$$

uniformly in the strip $1-\beta\leqslant\sigma\leqslant\beta$, provided that $\rho>2\beta-2$. We have only to note that $\zeta(s)=O(|t|^c)$ for some c, as $|t|\to\infty$, for $1-\beta\leqslant\sigma\leqslant\beta$, while $\dfrac{1}{\Gamma(s)}$ is an entire function of order 1. Hence

$$D^\rho(x)=Q_\rho(x) + \frac{1}{2\pi i}\int\limits_L \frac{\Gamma(s)\zeta^2(s)x^{s+\rho}}{\Gamma(\rho+1+s)}\,ds, \tag{14}$$

where

$$Q_\rho(x) = \frac{1}{2\pi i}\int\limits_{\mathscr{C}} \frac{\Gamma(s)\zeta^2(s)x^{s+\rho}}{\Gamma(\rho+1+s)}\,ds, \quad x>0, \tag{15}$$

$\mathscr{C}$ being the rectangle with vertices at $\beta-iR, \beta+iR, 1-r+iR, 1-r-iR$, in that order.

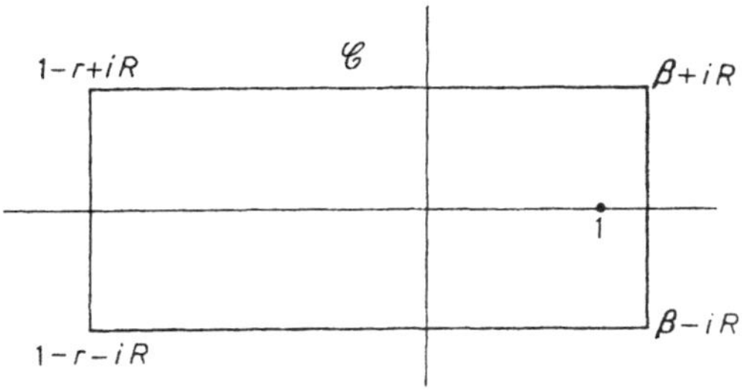

If we use equation (8) again in (14), and change the variable $s\to1-s$, we get

$$D^\rho(x)=Q_\rho(x) + \frac{1}{2\pi i}\int\limits_{\mathscr{C}_{\beta,r}} \frac{\pi^{-2s+1}\Gamma(1-s)\Gamma^2(\frac{1}{2}s)\zeta^2(s)x^{1-s+\rho}}{\Gamma(\rho+2-s)\Gamma^2(\frac{1}{2}-\frac{1}{2}s)}\,ds, \tag{16}$$

with the notation that, *for any real* α, $\mathscr{C}_{\alpha,r}$ denotes the broken line through $\alpha-i\infty, \alpha-iR, r-iR, r+iR, \alpha+iR$, and $\alpha+i\infty$, in that order.

Hence, if $\beta\geqslant1+\delta>1$, and $\rho>2\beta-1$, we have

$$D^\rho(x)-Q_\rho(x) = \frac{1}{2\pi i}\int\limits_{\mathscr{C}_{\beta,r}} \frac{\pi^{-2s+1}\Gamma(1-s)\Gamma^2(\frac{1}{2}s)\zeta^2(s)x^{1-s+\rho}}{\Gamma(\rho+2-s)\Gamma^2(\frac{1}{2}-\frac{1}{2}s)}\,ds, \tag{17}$$

the integral converging *absolutely*.

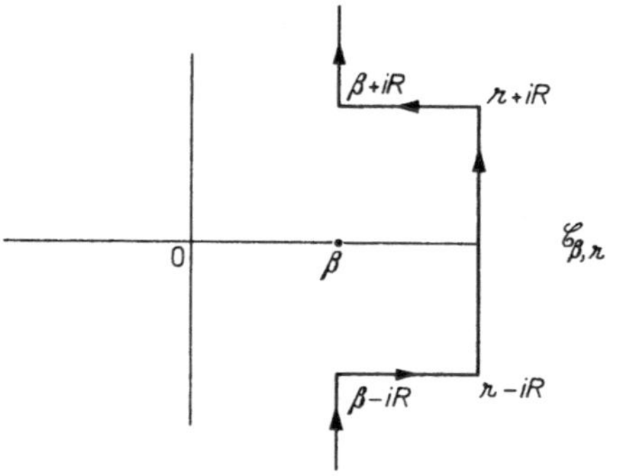

§ 3. **An auxiliary function.** Formula (17) suggests a study of the function f_ρ *defined by*

$$f_\rho(x) = \frac{1}{2\pi i} \int\limits_{\mathscr{C}_{\alpha,r}} G_\rho(s) x^{1+\rho-s} ds, \qquad G_\rho(s) = \frac{\Gamma(1-s)\Gamma^2(\frac{1}{2}s)}{\Gamma(\rho+2-s)\Gamma^2(\frac{1}{2}-\frac{1}{2}s)}, \tag{18}$$

for $x>0$. By formula (10), this integral converges absolutely for $\alpha<\frac{1}{2}(\rho+1)$, and is independent of α. We take $\alpha=-\frac{1}{3}$, and write $\mathscr{C}_r=\mathscr{C}_{-\frac{1}{3},r}$. Thus, for $x>0$, $\rho>-\frac{5}{3}$, we have

$$f_\rho(x) = \frac{1}{2\pi i} \int\limits_{\mathscr{C}_r} G_\rho(s) x^{1+\rho-s} ds, \qquad \mathscr{C}_r=\mathscr{C}_{-\frac{1}{3},r}. \tag{19}$$

It is immediate that if f'_ρ denotes the derivative of f_ρ, with respect to x, then

$$f'_{\rho+1}(x)=f_\rho(x), \qquad x>0, \qquad \rho>-\tfrac{5}{3}. \tag{20}$$

The asymptotic behaviour of $f_\rho(x)$, as $x\to\infty$, is given by the formula

$$f_\rho(x)=2^{-\rho}(2\pi)^{-\frac{1}{2}} x^{\frac{1}{2}\rho+\frac{1}{4}} \cos(4x^{\frac{1}{2}}-\tfrac{1}{2}\rho\pi-\tfrac{1}{4}\pi)+O(x^{\frac{1}{2}\rho+\varepsilon}), \tag{21}$$
$$0<\varepsilon<\tfrac{1}{4}, \qquad \rho>-\tfrac{5}{3}.$$

In order to prove this, we shall replace $G_\rho(s)$ in (19) by

$$F_\rho(s)=2^{-(\rho+\frac{1}{2})}\pi^{-\frac{1}{2}} \cdot \frac{2\pi(4)^{-2(s-\frac{3}{4}-\frac{\rho}{2})}\Gamma(2s-\frac{3}{2}-\rho)}{\Gamma(-s+\frac{3}{2}+\rho)\Gamma(s-\frac{1}{2}-\rho)}. \tag{22}$$

Since $\Gamma(s)\Gamma(1-s)=\pi/(\sin\pi s)$,

$$F_\rho(s)=2^{-(\rho+\frac{1}{2})}\pi^{-\frac{1}{2}}\cdot 2(4)^{-(2s-\frac{3}{2}-\rho)}\Gamma(2s-\tfrac{3}{2}-\rho)\cos[\pi(s-\rho-1)]. \tag{23}$$

By Stirling's formula,

$$\log\Gamma(z+\alpha)=(z+\alpha-\tfrac{1}{2})\log z-z+\tfrac{1}{2}\log 2\pi+O\left(\frac{1}{|z|}\right), \tag{24}$$

as $|z|\to\infty$, uniformly for $-\pi+\delta\leqslant\arg z\leqslant\pi-\delta$, for any $\delta>0$. Applying this to the gamma factors in $G_\rho(s)$ and $F_\rho(s)$, we get

$$\log G_\rho(s)-\log F_\rho(s)=O\left(\frac{1}{|s|}\right), \quad \text{as} \quad |s|\to\infty,$$

uniformly for $0<\delta<|\arg s|<\pi-\delta$, for every $\delta>0$, so that

$$G_\rho(s)-F_\rho(s)=F_\rho(s)\cdot O\left(\frac{1}{|s|}\right). \tag{25}$$

Hence

$$f_\rho(x)=\frac{1}{2\pi i}\int_{\mathscr{C}_r} F_\rho(s)x^{1+\rho-s}ds+\frac{1}{2\pi i}\int_{\mathscr{C}_r} F_\rho(s)\cdot O\left(\frac{1}{|s|}\right)\cdot x^{1+\rho-s}ds. \tag{26}$$

The integrand in the second integral on the right is analytic, and we can change the path of integration from $\mathscr{C}_r=\mathscr{C}_{-\frac{1}{3},r}$ to $\mathscr{C}_{1+\frac{1}{2}\rho-\varepsilon,r}$, $0<\varepsilon<\frac{1}{4}$. Hence

$$\frac{1}{2\pi i}\int_{\mathscr{C}_r} F_\rho(s)\cdot O\left(\frac{1}{|s|}\right)\cdot x^{1+\rho-s}ds=\frac{1}{2\pi i}\int_{\mathscr{C}_{1+\frac{1}{2}\rho-\varepsilon,r}} =O(x^{\frac{1}{2}\rho+\varepsilon}). \tag{27}$$

We evaluate the first integral on the right-hand side of (26) by means of the formula

$$\cos(x+a)=\frac{1}{2\pi i}\int_{\mathscr{C}_{-\frac{1}{3},r}} \Gamma(s)\cos(a+\tfrac{1}{2}\pi s)x^{-s}ds, \tag{28}$$

where a is real, and $x>0$. This can be proved by contour integration. We have only to note that if we expand $\cos(x+a)$ in a neighbourhood of the point a, then the left-hand side of (28) gives the sum of the residues of the integrand at its poles. From (23) and (28) we obtain

$$\frac{1}{2\pi i}\int_{\mathscr{C}_r} F_\rho(s)x^{1+\rho-s}ds=2^{-\rho}(2\pi)^{-\frac{1}{2}}x^{\frac{1}{2}\rho+\frac{1}{4}}\cos(4x^{\frac{1}{2}}-\tfrac{1}{2}\rho\pi-\tfrac{1}{4}\pi). \tag{29}$$

From (29), (27) and (26) we obtain (21).

We can take more terms in Stirling's formula, and have

$$\log\Gamma(z+\alpha)=(z+\alpha-\tfrac{1}{2})\log z-z+\tfrac{1}{2}\log 2\pi+\sum_{\nu=1}^{m} c_\nu z^{-\nu}+O\left(\frac{1}{|z|^{m+1}}\right), \quad (24)'$$

where m is a positive integer, and $c_1=A_0+\frac{1}{2}\alpha(\alpha-1)$, where A_0 is an absolute constant. We then have

$$G_\rho(s)=F_{\rho,0}(s)\left[1+\sum_{\nu=1}^{m}\frac{c'_\nu}{(2s)^\nu}+O\left(\frac{1}{|s|^{m+1}}\right)\right]$$

$$=\sum_{\nu=0}^{m} F_{\rho,\nu}(s)+F_{\rho,0}(s)\cdot O\left(\frac{1}{|s|^{m+1}}\right), \qquad F_{\rho,0}(s)=F_\rho(s),$$

where

$$F_{\rho,\nu}(s)=\frac{4^{-2(s-\frac{3}{4}-\frac{1}{2}\rho)}c''_\nu\,\Gamma(2s-\frac{3}{2}-\rho-\nu)}{\Gamma(-s+\frac{3}{2}+\rho)\,\Gamma(s-\frac{1}{2}-\rho)},$$

and $\int F_{\rho,\nu}(s)x^{-s}ds$ can be evaluated as in the case $\nu=0$. We thus obtain a stronger formula than (21):

$$f_\rho(x)=\sum_{\nu=0}^{m}\kappa_\nu x^{\frac{1}{2}(\rho+\frac{1}{2}-\nu)}\cos(4x^{\frac{1}{2}}-\tfrac{1}{2}\rho\pi-\tfrac{1}{4}\pi+\tfrac{1}{2}\nu\pi)+O(x^{\frac{1}{2}(\rho+\frac{1}{2}-m-1)}), \quad (30)$$

where m is a positive integer, $\rho>-\frac{5}{3}$, and the κ_ν are independent of x.

Formula (21) is enough to see that

$$\sum_{n=1}^{m} d(n)n^{-1-\rho}f_\rho(nx), \qquad x>0, \quad (31)$$

converges absolutely, and uniformly in every finite x-interval, for any $\rho>\frac{1}{2}$.

§ 4. **An identity involving the divisor function.** The absolute convergence of the series in (31) can be used to prove the following

THEOREM 1. *If* $x>0$, *and* $\rho>\frac{1}{2}$, *we have the identity*

$$D^\rho(x)-Q_\rho(x)=W_\rho(x), \quad (32)$$

where

$$D^\rho(x)=\frac{1}{\Gamma(\rho+1)}\sum_{n\leqslant x} d(n)\,(x-n)^\rho, \quad (33)$$

$$W_\rho(x)=\pi\sum_{n=1}^{\infty} d(n)\,(\pi^2 n)^{-1-\rho}f_\rho(\pi^2 nx), \quad (34)$$

and

$$Q_\rho(x) = \frac{1}{2\pi i}\int_{\mathscr{C}} \frac{\Gamma(s)\zeta^2(s)x^{s+\rho}}{\Gamma(\rho+1+s)}\,ds, \tag{35}$$

where $\mathscr{C}$ *is a rectangle with vertices at* $\beta - iR, \beta + iR, 1-r+iR, 1-r-iR$, R *and* r *being suitably chosen, fixed, positive numbers,* $r>\rho+1$, *and* β *any fixed number greater than* 1. (*If* ρ *is a positive integer,* $\mathscr{C}$ *encloses all the poles of the integrand*).

PROOF. We can substitute the series $\sum_{n=1}^{\infty} d(n)n^{-s}$ for $\zeta^2(s)$ in (17), provided that $\operatorname{Re} s > 1$, and interchange the summation and integration, provided that

$$\int_{\mathscr{C}_{\beta,r}} \sum_{n=1}^{\infty} \frac{d(n)}{|n^s|}\cdot\frac{|\Gamma(1-s)\Gamma^2(\frac{1}{2}s)x^{1-s+\rho}|}{|\Gamma(\rho+2-s)\Gamma^2(\frac{1}{2}-\frac{1}{2}s)|}\cdot|ds|<\infty,$$

which is the case if $\rho > 2\beta - 1$. Hence

$$D^\rho(x) - Q_\rho(x) = W_\rho(x), \quad \text{for} \quad x>0, \quad \rho>1. \tag{36}$$

This identity remains unchanged in form by differentiation because of (20), provided that the series on the right-hand side which results from the differentiation is absolutely convergent. But this is so for $\rho > \frac{1}{2}$, because of (31).

COROLLARY. *If* $x>0$, *and* ρ *a positive integer, then*

$$D^\rho(x) - Q_\rho(x) = -(2\pi)^{-\rho}\sum_{n=1}^{\infty} d(n)\left(\frac{x}{n}\right)^{\frac{1}{2}(1+\rho)} H_{1+\rho}\{4\pi(nx)^{\frac{1}{2}}\}, \tag{37}$$

where

$$H_\nu(x) = Y_\nu(x) + (-1)^{\nu-1}\left(\frac{2}{\pi}\right)K_\nu(x), \tag{38}$$

Y_ν *and* K_ν *being the Bessel functions of the second kind.*

The proof results from the following formula, which can be obtained by contour integration: if $x>0$, and k a positive, even integer, then

$$\frac{1}{2\pi i}\int_{\mathscr{C}_{\beta,r}} \frac{\Gamma^2(\frac{1}{2}s)\Gamma(1-s)x^{k-s}}{\Gamma^2(\frac{1}{2}-\frac{1}{2}s)\Gamma(k+1-s)}\,ds = -2^{1-k}x^{\frac{1}{2}k}H_k(4\sqrt{x}). \tag{39}$$

This enables us to deduce (37) from (32) if ρ is a positive, odd integer, and because

$$\frac{d}{dx}(x^{\frac{1}{2}\nu}H_\nu\{4\pi(nx)^{\frac{1}{2}}\}) = 2\pi n^{\frac{1}{2}}\cdot x^{\frac{1}{2}(\nu-1)}H_{\nu-1}\{4\pi(nx)^{\frac{1}{2}}\},$$

(37) holds also for any positive, even integer, by integration.

§ 5. Voronoi's theorem. Identity (32), with $\rho=1$, can be used to prove the following

THEOREM 2 (G. VORONOI). *If* $E(x)$ *is defined as in* (1), *then*

$$E(x)=O(x^{\frac{1}{3}}\log x), \quad as \quad x\to\infty. \tag{40}$$

PROOF. Let the difference operator Δ_y be defined by

$$\Delta_y f(x) = f(x+y) - f(x), \qquad 0<y<x,$$

for any function f defined in $(0,\infty)$. If f is differentiable, then

$$\Delta_y f(x) = \int_x^{x+y} f'(t)\,dt.$$

If we apply the operator Δ_y to the identity (32), with $\rho=1$, we obtain

$$\Delta_y D^1(x) - \Delta_y Q_1(x) = \Delta_y W_1(x). \tag{41}$$

Now

$$\Delta_y D^1(x) = \Delta_y\Big(\sum_{n\leqslant x} d(n)(x-n)\Big) = y\sum_{n\leqslant x} d(n) + \sum_{x<n\leqslant x+y} d(n)(x+y-n),$$

while

$$\Delta_y Q_1(x) = \int_x^{x+y} Q_0(u)\,du,$$

where

$$Q_0(x) = \frac{1}{2\pi i}\int_{\mathscr{C}} \frac{\zeta^2(s)x^s}{s}\,ds,$$

$\mathscr{C}$ being a curve which encloses all the poles of the integrand. Since

$$\zeta(s) = \frac{1}{s-1} + \gamma + O(|s-1|), \tag{42}$$

in a neighbourhood of $s=1$, where γ is Euler's constant (though it is sufficient here to know that γ is some constant), and $\zeta(0)=-\frac{1}{2}$, we see that

$$Q_0(x) = \tfrac{1}{4} + (2\gamma-1)x + x\log x.$$

Hence

$$\Delta_y Q_1(x) = y\,Q_0(x) + O(y^2\log x). \tag{43}$$

Thus

$$\Delta_y(D^1(x) - Q_1(x)) \geqslant y(D_0(x) - Q_0(x)) + O(y^2\log x). \tag{44}$$

Further

$$\Delta_y f_1(x) = \begin{cases} O\Big(\sup\limits_{x\leqslant u\leqslant x+y} |f_1(u)|\Big) = O(x^{\frac{3}{4}}), \\ O\Big(y \sup\limits_{x\leqslant u\leqslant x+y} |f_1'(u)|\Big) = O(y x^{\frac{1}{4}}), \end{cases} \tag{45}$$

since $f_1'(x) = f_0(x) = O(x^{\frac{1}{4}})$, as $x \to \infty$, by (20) and (21). Thus

$$\Delta_y f_1(nx) = O(x^{\frac{1}{4}} n^{\frac{3}{4}} \min(x^{\frac{1}{2}}, n^{\frac{1}{2}} y)),$$

since $(f_1(nx))' = nf_0(nx)$. Hence

$$\Delta_y \left(\sum_{n=1}^{\infty} \frac{d(n)}{n^2} f_1(\pi^2 nx) \right) = O\left(\sum_{n \leqslant z} \frac{d(n)}{n^{\frac{3}{4}}} \cdot y x^{\frac{1}{4}} \right) + O\left(\sum_{n > z} \frac{d(n)}{n^{\frac{5}{4}}} \cdot x^{\frac{3}{4}} \right),$$

where $z = x^{\frac{1}{3}}$, $x > 1$. Since $D^0(x) = O(x \log x)$, by (2), we obtain by partial summation,

$$\sum_{n \leqslant z} \frac{d(n)}{n^{\frac{3}{4}}} \cdot y x^{\frac{1}{4}} = O(z^{\frac{1}{4}} \log z) \cdot y x^{\frac{1}{4}},$$

$$\sum_{n > z} \frac{d(n)}{n^{\frac{5}{4}}} \cdot x^{\frac{3}{4}} = O(z^{-\frac{1}{4}} \log z) \cdot x^{\frac{3}{4}}.$$

Hence

$$\Delta_y \left(\sum_{n=1}^{\infty} \frac{d(n)}{n^2} f_1(\pi^2 nx) \right) = O(y x^{\frac{1}{4}} z^{\frac{1}{4}} \log z) + O(x^{\frac{3}{4}} z^{-\frac{1}{4}} \log z).$$

If we choose $y = x^{\frac{1}{3}}$, $z = x^{\frac{1}{3}}$, we get

$$\Delta_y \left(\sum_{n=1}^{\infty} \frac{d(n)}{n^2} f_1(\pi^2 nx) \right) = O(x^{\frac{2}{3}} \log x). \tag{46}$$

From (44) we have

$$\Delta_y (D^1(x) - Q_1(x)) \geqslant x^{\frac{1}{3}} (D^0(x) - Q_0(x)) + O(x^{\frac{2}{3}} \log x),$$

and from (46) we obtain

$$\Delta_y (D^1(x) - Q_1(x)) = O(x^{\frac{2}{3}} \log x). \tag{47}$$

Hence

$$D^0(x) - Q_0(x) \leqslant O(x^{\frac{1}{3}} \log x). \tag{48}$$

Since $d(n) > 0$, $D^0(x)$ is monotone, and since

$$\Delta_y D^1(x) = \int_x^{x+y} D^0(u)\, du,$$

we have

$$y D^0(x+y) \geqslant \Delta_y D^1(x) \geqslant y D^0(x). \tag{49}$$

From (47), (49), and (43), it follows that

$$D^0(x+y)-Q_0(x)\geqslant O(x^{\frac{1}{3}}\log x), \qquad y=x^{\frac{1}{3}},$$

while $Q_0(x+y)-Q_0(x)=O(x^{\frac{1}{3}}\log x)$, by the definition of Q_0. Hence $D^0(x)-Q_0(x)\geqslant O(x^{\frac{1}{3}}\log x)$, which, taken together with (48), gives the result

$$D^0(x)-Q_0(x)=O(x^{\frac{1}{3}}\log x). \tag{50}$$

This implies the theorem, since $Q_0(x)=\frac{1}{4}+(2\gamma-1)x+x\log x$. By Dirichlet's estimate (2), it follows that the constant γ in (42) must be Euler's constant.

§ 6. A theorem of A. S. Besicovitch. For the proof of (4) and (5) we require the result that the square-roots of positive, square-free integers are linearly independent over the field of rational numbers. This is a particular case of a general theorem, proved by Besicovitch, that a polynomial $P(p_1^{1/n_1},\ldots,p_r^{1/n_r})$ in $p_1^{1/n_1},\ldots,p_r^{1/n_r}$, of degree less than or equal to n_1-1 in p_1^{1/n_1}, less than or equal to n_2-1 in p_1^{1/n_2}, and so on, with rational coefficients, not all zero, cannot vanish. Here $p_1,\ldots,p_r$ are different primes. But the special case we require is a consequence of the following

LEMMA. *Let Q be the field of rational numbers. Let $a_j\in Q$, for $j=1,2,\ldots,r$. Let $q_j=a_jp_j$, where $p_1,p_2,\ldots p_r$ are different primes. We suppose that both the numerator and denominator of a_j are prime to $p_1,\ldots,p_r$. Then $\sqrt{q_j}$ does not lie in the field $Q_j=Q(\sqrt{q_1},\ldots,\sqrt{q_{j-1}},\sqrt{q_{j+1}},\ldots,\sqrt{q_r})$.*

PROOF. The case $r=1$ is easily seen to be true. The general case can be proved by induction on r. We suppose the lemma true for r and prove it for $r+1$.

Let $n=q_{r+1}$, $m=q_r$. Suppose that

$$\sqrt{n}=c_0+c_1\sqrt{m}, \quad \text{where} \quad c_0,c_1\in Q(\sqrt{q_1},\ldots,\sqrt{q_{r-1}})\equiv\mathscr{F}, \text{ say.}$$

Then $2c_0c_1\sqrt{m}=n-c_0^2-c_1^2m\in\mathscr{F}$. By the induction hypothesis, $c_0c_1=0$. If $c_1=0$, then $\sqrt{n}\in\mathscr{F}$, which is impossible. Hence $c_0=0$, or $\sqrt{\dfrac{n}{m}}=c_1\in\mathscr{F}$, which is not possible, by the induction hypothesis applied to $\left(\sqrt{q_1},\ldots,\sqrt{q_{r-1}},\sqrt{\dfrac{n}{m}}\right)$.

§ 7. Theorems of Hardy and of Ingham. We shall use the above lemma, as well as the identity (32) with $\rho=1$, to prove the following

THEOREM 3. *If* $E(x)$ *is defined as in* (1), *then*

(G. H. Hardy)
$$\limsup_{x\to\infty} \frac{E(x)}{x^{\frac{1}{4}}} = +\infty, \tag{4}$$

and

(A. E. Ingham)
$$\liminf_{x\to\infty} \frac{E(x)}{x^{\frac{1}{4}}} = -\infty. \tag{5}$$

PROOF. We start from the identity

$$D^1(x) - Q_1(x) = \pi \sum_{n=1}^{\infty} d(n)(\pi^2 n)^{-2} f_1(\pi^2 n x), \tag{51}$$

which is the case $\rho=1$ of (32). Here $D^1(x)$ is defined as in (6), and

$$Q_1(x) = \frac{1}{2\pi i} \int_{\mathscr{C}} \frac{\Gamma(s)\zeta^2(s) x^{1+s}}{\Gamma(2+s)}\, ds, \tag{52}$$

where $\mathscr{C}$ is a curve which encloses all the poles of the integrand. Using the asymptotic formula for $f_1(\pi^2 n x)$ given by (21), with $0<\varepsilon<\frac{1}{8}$, we have

$$D^1(x) - Q_1(x) = c \sum_{n=1}^{\infty} \frac{d(n)}{n^2} (nx)^{\frac{3}{4}} \cos(4\pi x^{\frac{1}{2}} n^{\frac{1}{2}} - \tfrac{3}{4}\pi) + O(x^{\frac{5}{8}}), \tag{53}$$

where c is a positive constant.

Let r be a positive integer, to be chosen sufficiently large, and let $\eta = r + \frac{3}{2}$.

If we take the relation (53), with x^2 in place of x, and multiply it throughout by

$$\sigma^\eta \left(\frac{d}{dx}\cdot\frac{1}{x}\right)(e^{-sx} x^r), \qquad \sigma = \operatorname{Re} s > 0,$$

and integrate with respect to x from 0 to ∞, the left-hand side of (53) gives

$$\sigma^\eta \int_0^\infty \left(\frac{d}{dx}\cdot\frac{1}{x}\right)(e^{-sx} x^r)\cdot(D^1(x^2) - Q_1(x^2))\, dx$$

$$= (-2)\sigma^\eta \int_0^\infty e^{-sx} x^r \{D^0(x^2) - Q_0(x^2)\}\, dx, \tag{54}$$

since $D^0(x)-Q_0(x)=O(x^{\frac{1}{2}})$ by (2). On the other hand,

$$\left(\frac{d}{dx}\cdot\frac{1}{x}\right)(e^{-sx}x^r)=e^{-sx}((r-1)x^{r-2}-sx^{r-1}),$$

so that the O-term on the right-hand side of (53) gives

$$\sigma^\eta\int_0^\infty\left(\frac{d}{dx}\cdot\frac{1}{x}\right)(e^{-sx}x^r)\cdot O(x^{\frac{5}{4}})\cdot dx=\sigma^\eta\cdot O\left(\frac{1}{\sigma^{r+\frac{1}{4}}}\right)+s\cdot\sigma^\eta\cdot O\left(\frac{1}{\sigma^{r+\frac{5}{4}}}\right). \quad (55)$$

The main term on the right-hand side of (53) leads us to consider, for fixed t, $t=\operatorname{Im} s$,

$$S(\sigma)=\sigma^\eta\int_0^\infty\left(\frac{d}{dx}\cdot\frac{1}{x}\right)(e^{-sx}x^r)\cdot\left[\sum_{n=1}^\infty\frac{d(n)}{n^2}\cdot n^{\frac{3}{4}}\cdot x^{\frac{3}{2}}\cos(4\pi x n^{\frac{1}{2}}-\tfrac{3}{4}\pi)\right]dx$$

$$=\sigma^\eta\int_0^\infty[(r-1)x^{r-2}-sx^{r-1}]x^{\frac{3}{2}}e^{-sx}\left[\sum_{n=1}^\infty\frac{d(n)}{n^{\frac{5}{4}}}\cos(4\pi x n^{\frac{1}{2}}-\tfrac{3}{4}\pi)\right]dx$$

$$=\sum_{n=1}^\infty\frac{d(n)}{n^{\frac{5}{4}}}\sum_{\nu=r-2}^{r-1}e_\nu s^{2-r+\nu}\cdot\sigma^\eta\int_0^\infty e^{-sx}x^{\nu+\frac{3}{2}}\cos(4\pi x n^{\frac{1}{2}}-\tfrac{3}{4}\pi)\,dx, \quad (56)$$

where $e_{r-2}=r-1$, and $e_{r-1}=-1$.

Now, for $0<\sigma<1$, and t fixed, we have

$$\left|\sigma^\eta\int_0^\infty e^{-sx}x^{\nu+\frac{3}{2}}\cos(4\pi x n^{\frac{1}{2}}-\tfrac{3}{4}\pi)\,dx\right|\leqslant c\cdot\sigma^{\eta-\nu-\frac{5}{2}},\ \nu=r-2,\ r-1,$$

where c is a constant, uniformly in n. Here $\eta-\nu-\frac{5}{2}\geqslant 0$, since $\eta=r+\frac{3}{2}$. Hence

$$\lim_{\sigma\to 0+}S(\sigma)$$

$$=\sum_{n=1}^\infty\frac{d(n)}{n^{\frac{5}{4}}}\lim_{\sigma\to 0+}\sum_{\nu=r-2}^{r-1}e_\nu s^{2-r+\nu}\cdot\sigma^\eta\int_0^\infty e^{-sx}x^{\nu+\frac{3}{2}}\cdot\cos(4\pi x n^{\frac{1}{2}}-\tfrac{3}{4}\pi)\,dx. \quad (57)$$

Now set

$$V(x)=1+\frac{1}{2}(e^{ix}+e^{-ix})=2\cos^2\frac{x}{2}=1+\cos x,$$

and

$$W(x)=\prod_{k=1}^N V(4\pi n_k^{\frac{1}{2}}x-\theta_k),$$

where N is a positive integer, θ_k a real number appropriately to be chosen, and (n_k), $k=1,2,\ldots$, is the sequence of positive, square-free, integers. The product W can be rewritten as

$$W(x)=1+\frac{1}{2}\sum_{k=1}^{N}\left[\exp(4\pi i n_k^{\frac{1}{2}}x-i\theta_k)+\exp(-4\pi i n_k^{\frac{1}{2}}x+i\theta_k)\right]+U(x),$$

where $U(x)$ is a trigonometric polynomial of the form $\sum_{\nu=1}^{M} b_\nu e^{ic_\nu x}$, where the b_ν are complex numbers, and the c_ν are real numbers of the form

$$4\pi(r_1 n_1^{\frac{1}{2}}+r_2 n_2^{\frac{1}{2}}+\cdots+r_N n_N^{\frac{1}{2}}), \tag{58}$$

where $r_\nu=0$, $+1$, or -1 and at least two of the r_ν's are different from zero. By Besicovitch's lemma, proved in § 6, none of the numbers in (58) equals $\pm 4\pi n^{\frac{1}{2}}$ for any integer n.

Let us now consider the limit

$$L=\lim_{\sigma\to 0+}\frac{\sigma^\eta}{\Gamma(\eta)}\int_0^\infty e^{-\sigma x}x^r(D^0(x^2)-Q_0(x^2))\cdot W(x)\,dx. \tag{59}$$

This requires that we consider (54) with $s=\sigma\pm 4\pi i n_k^{\frac{1}{2}}$, and take the limit as $\sigma\to 0+$; or, equivalently, consider (55) and (56) with $s=\sigma\pm 4\pi i n_k^{\frac{1}{2}}$ and take the limit as $\sigma\to 0+$. Clearly (55) contributes zero. To consider the contribution from (56), we note that if A and B are real, $A\neq 0$, $q>0$, $\operatorname{Re}s>0$, then

$$\int_0^\infty e^{-sx}x^q\cos(Ax+B)\,dx=\frac{1}{2}\left\{\frac{e^{iB}\Gamma(q+1)}{(s-iA)^{q+1}}+\frac{e^{-iB}\Gamma(q+1)}{(s+iA)^{q+1}}\right\}, \tag{60}$$

since

$$\int_0^\infty e^{-st}t^{\nu-1}\sin(\alpha t)\,dt=\frac{\Gamma(\nu)}{2i}\{(s-i\alpha)^{-\nu}-(s+i\alpha)^{-\nu}\},$$

and

$$\int_0^\infty e^{-st}t^{\nu-1}\cos(\alpha t)\,dt=\frac{\Gamma(\nu)}{2}\{(s+i\alpha)^{-\nu}+(s-i\alpha)^{-\nu}\},$$

for $\operatorname{Re}\nu>0$, $\operatorname{Re}s>|\operatorname{Im}\alpha|$. Hence

$$\lim_{\sigma\to 0+}\sigma^{q+1}\int_0^\infty e^{-sx}x^q\cos(Ax+B)\,dx=\begin{cases}0, & \text{if } s\neq\sigma\pm iA,\\ \frac{1}{2}\Gamma(q+1)e^{\pm iB}, & \text{if } s=\sigma\pm iA.\end{cases} \tag{61}$$

If we take $A=4\pi n^{\frac{1}{2}}$, $s=\sigma\pm 4\pi i n_k^{\frac{1}{2}}$, and use (61) in (57), we see that the only contribution is from the term corresponding to $\nu=r-1$.

Actually that term is

$$\sum_{n=1}^{\infty} \frac{d(n)}{n^{\frac{5}{4}}} \lim_{\sigma\to 0+} \left[(-1)s\cdot\sigma^{\eta} \int_0^{\infty} e^{-sx} x^{r+\frac{1}{2}} \cos(4\pi x n^{\frac{1}{2}} - \tfrac{3}{4}\pi)\,dx\right],$$

with $s=\sigma\pm 4\pi i n_k^{\frac{1}{2}}$. (Note that the part $1+U(x)$ of $W(x)$ contributes nothing, since $s\neq\sigma\pm iA$ in that polynomial). Thus we obtain

$$L=c_1\cdot\sum_{k=1}^{N} \frac{d(n_k)}{n_k^{\frac{3}{4}}}\sin(\tfrac{3}{4}\pi-\theta_k), \qquad c_1>0, \tag{62}$$

where L is defined as in (57). Since $W(x)\geqslant 0$, we have

$$\lim_{\sigma\to 0+}\frac{\sigma^{\eta}}{\Gamma(\eta)}\int_0^{\infty}[D^0(x^2)-Q_0(x^2)]e^{-\sigma x}x^r W(x)\,dx$$

$$\leqslant \limsup_{x\to\infty}\frac{D^0(x^2)-Q_0(x^2)}{x^{\frac{1}{2}}}\cdot\lim_{\sigma\to 0+}\frac{\sigma^{\eta}}{\Gamma(\eta)}\int_0^{\infty}e^{-\sigma x}x^{\eta-1}W(x)\,dx. \tag{63}$$

We have, however,

$$\lim_{\sigma\to 0+}\frac{\sigma^{\eta}}{\Gamma(\eta)}\int_0^{\infty}e^{-\sigma x}x^{\eta-1}W(x)\,dx=1,$$

since $W(x)$ is of the form $1+\sum\beta_\nu e^{i\gamma_\nu x}$, $\gamma_\nu\neq 0$. Hence (62) and (63) yield the inequality

$$\limsup_{x\to\infty}\frac{D^0(x^2)-Q_0(x^2)}{x^{\frac{1}{2}}}\geqslant c_1\sum_{k=1}^{N}\frac{d(n_k)}{n_k^{\frac{3}{4}}}\sin(\tfrac{3}{4}\pi-\theta_k).$$

If we choose $\theta_k=\frac{1}{4}\pi$, then, since this holds for any integer $N>0$,

$$\limsup_{x\to\infty}\frac{D^0(x)-Q_0(x)}{x^{\frac{1}{4}}}=+\infty.$$

Similarly, by choosing $\theta_k=\frac{5}{4}\pi$, we obtain

$$\liminf_{x\to\infty}\frac{D^0(x)-Q_0(x)}{x^{\frac{1}{4}}}=-\infty.$$

Since $Q_0(x)=\frac{1}{4}+(2\gamma-1)x+x\log x$, the theorem follows.

§ 8. Equiconvergence theorems of A. Zygmund. We shall prove that identity (32) holds for $\rho > -\frac{1}{2}$, provided that $x > 0$, x non-integral, by the application of a general theorem (Theorem 6) of A. Zygmund on the convergence of trigonometric integrals.

Given two trigonometric series $\sum_{n=-\infty}^{\infty} a_n e^{inx}$, $\sum_{n=-\infty}^{\infty} b_n e^{inx}$, (convergent or not) in an interval $a < x < b$, we say that they are (*uniformly*) *equiconvergent*, if the difference $\sum_{r=-n}^{n} \left\{a_r e^{irx} - b_r e^{irx}\right\}$ converges (uniformly) in that interval, as $n \to +\infty$. If this difference converges (uniformly) to zero, the given series are said to be (uniformly) equiconvergent *in the strict sense*.

Given a trigonometric series $\sum_{n=-\infty}^{\infty} c_n e^{inx}$, *with* $c_n \to 0$ as $|n| \to \infty$, let

$$F(x) = \frac{1}{2} c_0 x^2 - \sum_{n=-\infty}^{\infty}{}' \frac{c_n}{n^2} e^{inx},$$

and let $L(x)$ be a periodic function, which is differentiable sufficiently often, with $L(x) = 1$ for $a' \leqslant x \leqslant b'$, and $L(x) = 0$ for $0 \leqslant x < a$ and $b < x \leqslant 2\pi$, where $a < a' < b' < b$. It is a classical result of Riemann that if we take the Fourier series of $F(x) \cdot L(x)$, and differentiate it formally twice, the resulting series is uniformly equiconvergent with $\sum_{n=-\infty}^{\infty} c_n e^{inx}$ in the interval $[a', b']$.

This result has been extended by Zygmund to trigonometric integrals of the form $\int_{-\infty}^{\infty} e^{itx} d\chi(t)$. The extension enables us to deduce the convergence of series of the form $\sum c_n e^{i\sqrt{n}x}$ from that of ordinary trigonometric series. Because of asymptotic formula (30), the result can be used to improve upon identity (32).

Let $\varphi(u)$ be a (complex-valued) function defined for $-\infty < u < +\infty$, which is of bounded variation in every finite interval, and let

$$\beta_\varphi(w) = \int_w^{w+1} |d\varphi(u)|. \tag{64}$$

We *assume* that

$$\beta_\varphi(w) \to 0, \quad \text{as} \quad |w| \to \infty. \tag{65}$$

We also *assume* that

$$\varphi(u) = 0, \qquad -\delta \leqslant u \leqslant \delta, \qquad 0 < \delta < 1. \tag{66}$$

By (65) we have

$$\int_0^w |d\varphi(u)| = o(w), \quad \text{as} \quad w \to \infty. \tag{67}$$

Let

$$\Phi(u) = \int_0^u \frac{1}{(it)^2} d\varphi(t). \tag{68}$$

Then Φ is of bounded variation in every finite interval, and

$$\beta_\Phi(w) = o(|w|^{-2}), \quad \text{as} \quad |w| \to \infty. \tag{69}$$

Let λ be a function belonging to the class $C^\infty(-\infty, \infty)$, of infinitely differentiable functions in $(-\infty, \infty)$, with compact support in an interval (a, b). (We shall, when required, make the further assumption that $\lambda(x) = 1$ for $x \in [a', b']$, where $-\infty < a < a' < b' < b < +\infty$).

We define the function μ by the relation

$$\lambda(x) = \int_{-\infty}^{\infty} e^{ixt} \mu(t)\, dt, \quad \text{or} \quad \mu(x) = \frac{1}{2\pi} \int_{-\infty}^{\infty} e^{-ixt} \lambda(t)\, dt. \tag{70}$$

Then $\mu \in C^\infty(-\infty, \infty)$, and $\mu^{(r)}$, the rth derivative of μ, satisfies the condition

$$\mu^{(r)}(t) = O\left(\frac{1}{1 + |t|^k}\right), \quad \text{as} \quad |t| \to \infty, \tag{71}$$

for *every* $k > 0$, $r \geqslant 0$.

Let

$$\psi(y) = \int_{-\infty}^{\infty} \varphi(y - u)\, \mu(u)\, du. \tag{72}$$

This integral is absolutely convergent, and $\psi \in C^\infty(-\infty, \infty)$.

Let

$$\Psi(y) = \int_{-\infty}^{\infty} \Phi(y - u)\, \mu(u)\, du = \int_{-\infty}^{\infty} \mu(y - u)\, \Phi(u)\, du, \tag{73}$$

where Φ and μ are defined as in (68) and (70). The last integral converges absolutely, and uniformly in y. Further

$$\Psi \in C^\infty(-\infty, \infty), \quad \Psi^{(k)}(y) = \int_{-\infty}^{\infty} \Phi(y - u)\, \mu^{(k)}(u)\, du, \quad \Psi^{(k)}(y) = O(1), \tag{74}$$

for any non-negative integer k.

Given two trigonometric integrals

$$A = \int_{-\infty}^{\infty} e^{ixt}\,dV(t), \quad B = \int_{-\infty}^{\infty} e^{ixt}\,dU(t), \tag{75}$$

where V is any function of bounded variation in every finite interval, and $U \in C^{\infty}(-\infty, \infty)$, with $U^{(r)}(x) = o\left(\frac{1}{1+|x|^k}\right)$, for any integers $k > 0$, $r > 0$, the *formal product* of A by B, written $A \cdot B$, is defined to be the integral

$$\int_{-\infty}^{\infty} e^{ixt}\,dW(t), \quad \text{where} \quad W(t) = \int_{-\infty}^{\infty} V(t-u)\,dU(u), \tag{76}$$

on the assumption that the integral defining $W(t)$ exists for all t in $(-\infty, \infty)$, and $W(t)$ is of bounded variation in every finite interval. Under suitable assumptions, the formal product $A \cdot B$ is the same as $B \cdot A$.

The *formal derivative* of the integral A is defined to be $\int_{-\infty}^{\infty} (ix) e^{ixt}\,dV(t)$.

We consider the integral $\int_{-\infty}^{\infty} e^{ixt}\,dV(t)$ *convergent*, if the integral $\int_{-\omega}^{\omega} e^{ixt}\,dV(t)$ has a finite limit as $\omega \to +\infty$.

THEOREM 4 (ZYGMUND). *Let* $\omega > 0$, and

$$\Delta_{\omega}(x) = \int_{-\omega}^{\omega} e^{ixu}\,d\psi(u) - \lambda(x) \int_{-\omega}^{\omega} e^{ixu}\,d\varphi(u). \tag{77}$$

Then

$$\Delta_{\omega}(x) \to 0, \quad \text{as} \quad \omega \to \infty, \tag{78}$$

uniformly for x *in* $(-\infty, \infty)$.

PROOF. By partial integration, the first integral in (77) equals

$$e^{ix\omega}\psi(\omega) - e^{-ix\omega}\psi(-\omega) - ix \int_{-\omega}^{\omega} e^{ixu}\psi(u)\,du$$

$$= \int_{-\infty}^{\infty} e^{ix(\omega-y)}\varphi(\omega-y)e^{ixy}\mu(y)\,dy - \int_{-\infty}^{\infty} e^{ix(-\omega-y)}\varphi(-\omega-y)e^{ixy}\mu(y)\,dy$$

$$-ix \int_{-\omega}^{\omega} du \int_{-\infty}^{\infty} e^{ix(u-y)}\varphi(u-y)e^{ixy}\mu(y)\,dy$$

$$= \int_{-\infty}^{\infty} \left\{ \int_{-\omega-y}^{\omega-y} e^{ixu}\,d\varphi(u) \right\} e^{ixy}\mu(y)\,dy.$$

The second integral in (77) equals (on substituting for λ),

$$\int_{-\infty}^{\infty} e^{ixy}\mu(y)\left\{\int_{-\omega}^{\omega} e^{ixu}\,d\varphi(u)\right\}dy.$$

Hence

$$\Delta_\omega(x)=\int_{-\infty}^{\infty}\left\{\int_{-\omega-y}^{-\omega} e^{ixu}\,d\varphi(u)+\int_{\omega}^{\omega-y} e^{ixu}\,d\varphi(u)\right\}e^{ixy}\mu(y)\,dy. \tag{79}$$

Now

$$\left|\int_{\omega}^{\omega-y} e^{ixu}\,d\varphi(u)\right|\leqslant\left|\int_{\omega}^{\omega-y}|d\varphi(u)|\right|=o(|y|+1), \tag{80}$$

as $\omega\to\infty$, the o being uniform in x for $-\infty<x<\infty$. For if $y\leqslant\frac{1}{2}\omega$, and $\omega\to\infty$, then $\omega-y\to\infty$, and (80) follows from (65). If $y\geqslant 2\omega$, then $\omega-y\to-\infty$, as $\omega\to\infty$, and, by (67), the second integral in (80) is $o(y-\omega)+o(\omega)=o(y)$. If $\frac{1}{2}\omega\leqslant y\leqslant 2\omega$, then the same integral is $o(\omega)=o(y)$. All the o's are uniform in x.

Likewise

$$\int_{-\omega}^{-\omega-y} e^{ixu}\,d\varphi(u)=o(|y|+1), \qquad \text{as} \quad \omega\to\infty,$$

uniformly in x, $-\infty<x<\infty$. Since

$$\int_{-\infty}^{\infty}(|y|+1)\cdot|\mu(y)|\cdot dy<\infty,$$

it follows that $\Delta_\omega(x)\to 0$ as $\omega\to\infty$, uniformly in x, $-\infty<x<\infty$, and the theorem is proved.

Remarks. The *formal product* of two trigonometric series $\sum_{n=-\infty}^{\infty} c_n e^{inx}$ and $\sum_{n=-\infty}^{\infty} \gamma_n e^{inx}$ is defined to be

$$\sum_{n=-\infty}^{\infty} C_n e^{inx}, \qquad C_n=\sum_{r=-\infty}^{\infty} c_r\gamma_{n-r},$$

where the series defining the C_n are assumed to converge absolutely. The *formal derivative of* $\sum_{n=-\infty}^{\infty} c_n e^{inx}$ is $\sum_{n=-\infty}^{\infty}(in)c_n e^{inx}$.

Theorem 4 is an analogue (not in the strongest possible form) for integrals of the following result first proved for series by A. Rajchman.

THEOREM 4′(RAJCHMAN). *Suppose that* $c_n \to 0$, *as* $|n| \to \infty$, $\sum_{n=-\infty}^{\infty} |n\gamma_n| < \infty$, *and that* $\sum_{n=-\infty}^{\infty} \gamma_n e^{inx}$ *converges to sum* $\lambda(x)$. *Then the two series*

$$\sum_{n=-\infty}^{\infty} C_n e^{inx}, \qquad \lambda(x) \sum_{n=-\infty}^{\infty} c_n e^{inx}$$

are uniformly equiconvergent in the strict sense. In particular, if $\lambda(x)=0$ *for x in a set E*, $\sum_{n=-\infty}^{\infty} C_n e^{inx}$ *converges uniformly to* 0 *in E*.

The condition $\sum_{n=-\infty}^{\infty} |n\gamma_n| < \infty$ is satisfied if $\sum \gamma_n e^{inx}$ is the expansion of a function which is differentiable sufficiently often.

Let us denote by I_1, I_2, I_3 the following trigonometric integrals:

$$I_1 = \int_{-\infty}^{\infty} e^{ixu} d\Phi(u), \quad I_2 = \int_{-\infty}^{\infty} e^{ixu} \mu(u)\, du, \quad I_3 = I_1 \cdot I_2 = \int_{-\infty}^{\infty} e^{ixu} d\Psi(u), \tag{81}$$

where Φ, μ, Ψ are defined as in (68), (70), and (73). Because of (69) and (71), the integrals I_1, I_2 converge absolutely. (We shall see later that I_3 also converges absolutely). We indicate by a dash the *formal derivatives* of these integrals. Thus

$$I_1' \equiv \int_{-\infty}^{\infty} iue^{ixu} d\Phi(u) = i \int_{-\infty}^{\infty} e^{ixu} d\Phi_1(u), \qquad \Phi_1(u) = \int_0^u t\, d\Phi(t), \tag{82}$$

$$I_2' \equiv \int_{-\infty}^{\infty} iue^{ixu} \mu(u) du = i \int_{-\infty}^{\infty} e^{ixu} d\mu_1(u), \qquad \mu_1(u) = \int_0^u t\mu(t)\, dt. \tag{83}$$

Again, by definition,

$$\begin{aligned} I_3' &\equiv \int_{-\infty}^{\infty} iue^{ixu} d\Psi(u) = i \int_{-\infty}^{\infty} e^{ixu} d\Psi_1(u), \\ \Psi_1(u) &= \int_0^u t\, d\Psi(t) = u\,\Psi(u) - \int_0^u \Psi(t)\, dt. \end{aligned} \tag{84}$$

If we indicate by a dot the operation of formal product, then, by definition,

$$I_1' \cdot I_2 + I_1 \cdot I_2' = i \int_{-\infty}^{\infty} e^{ixu} d\xi(u) + i \int_{-\infty}^{\infty} e^{ixu} d\eta(u),$$

where

$$\xi(u) = \int_{-\infty}^{\infty} \Phi_1(u-v)\mu(v)\, dv, \qquad \eta(u) = \int_{-\infty}^{\infty} \Phi(u-v)\, v\, \mu(v)\, dv, \tag{85}$$

the last two integrals converging absolutely. Now

$$\xi(u)-\int_{-\infty}^{\infty}\Phi_1(-v)\mu(v)\,dv+\eta(u)-\int_{-\infty}^{\infty}\Phi(-v)v\mu(v)\,dv$$

$$=\int_{-\infty}^{\infty}\left(\int_{-v}^{u-v}t\,d\Phi(t)\right)\mu(v)\,dv+\int_{-\infty}^{\infty}\left(\int_{-v}^{u-v}v\,d\Phi(t)\right)\mu(v)\,dv$$

$$=\int_{-\infty}^{\infty}\left(\int_{-v}^{u-v}(t+v)\,d\Phi(t)\right)\mu(v)\,dv$$

$$=\int_{-\infty}^{\infty}\left(\int_{0}^{u}t\,d\Phi(t-v)\right)\mu(v)\,dv$$

$$=u\Psi(u)-\int_0^u\Psi(t)\,dt, \tag{86}$$

since the interchange of the order of integration at the last stage is permitted. From (84) and (86) we see that $\xi(u)+\eta(u)$ differs from $\Psi_1(u)$ by a constant. Hence we obtain the relation

$$I_3'=I_1'\cdot I_2+I_1\cdot I_2'. \tag{87}$$

By repetition of the rule, we get

$$I_3''=I_1''\cdot I_2+2I_1'\cdot I_2'+I_1\cdot I_2'', \tag{88}$$

all the products having a meaning because of our assumptions.

REMARK. The analogue of (88) for trigonometric series is as follows. If S_1 and S_2 are trigonometric series, then

$$(S_1\cdot S_2)''=S_1''\cdot S_2+2S_1'\cdot S_2'+S_1\cdot S_2'', \tag{89}$$

where a dash indicates formal differentiation, and a dot indicates formal product, assuming that all the formal products exist.

Rule (88) is used in the proof of the following

THEOREM 5 (ZYGMUND). *Let φ and λ be defined as above, and let λ satisfy the additional condition that $\lambda(x)=1$ for $x\in[a',b']$, $a<a'<b'<b$. Let*

$$F(x)=\int_{-\infty}^{\infty}e^{ixu}\,d\Phi(u),\quad (=I_1), \tag{90}$$

the integral converging absolutely because of (69). If $\omega>0$, we then have

$$\int_{-\omega}^{\omega}e^{ixu}\,d\varphi(u)-\frac{1}{\pi}\int_a^b F(t)\lambda(t)\frac{d^2}{dt^2}\left(\frac{\sin\omega(x-t)}{x-t}\right)dt\to 0, \tag{91}$$

as $\omega\to\infty$, uniformly for $a'\leqslant x\leqslant b'$.

PROOF. We have, by (82) and (83),

$$I_1' \cdot I_2' = i^2 \int_{-\infty}^{\infty} e^{ixu} d\chi(u), \qquad \chi(u) = \int_{-\infty}^{\infty} \Phi_1(u-v)\, v\, \mu(v)\, dv, \tag{92}$$

while, by (70),

$$\lambda'(x) = \int_{-\infty}^{\infty} iu e^{ixu} \mu(u)\, du = i \int_{-\infty}^{\infty} e^{ixu} d\mu_1(u), \qquad \mu_1(u) = \int_0^u t\,\mu(t)\, dt, \tag{93}$$

where λ' denotes the ordinary derivative of λ.

By Theorem 4, with $i\chi$, λ', Φ_1 in place of ψ, λ, φ respectively, and with $\mu(v)$ replaced by $iv\mu(v)$, we have

$$i \int_{-\omega}^{\omega} e^{ixu} d\chi(u) - \lambda'(x) \int_{-\omega}^{\omega} e^{ixu} d\Phi_1(u) \to 0, \tag{94}$$

as $\omega \to \infty$, uniformly in x. Since $\lambda'(x) = 0$ for $x \in [a', b']$, it follows that

$$\int_{-\omega}^{\omega} e^{ixu} d\chi(u) \to 0, \quad \text{as} \quad \omega \to \infty, \tag{95}$$

uniformly for $a' \leqslant x \leqslant b'$; similarly also

$$\int_{-\omega}^{\omega} e^{ixu} d\chi^*(u) \to 0, \tag{96}$$

where

$$\chi^*(u) = \int_0^u \Phi(u-v)\, d\mu_2(v), \qquad \mu_2(v) = \int_0^v t^2 \mu(t)\, dt, \tag{97}$$

uniformly for $a' \leqslant x \leqslant b'$, since $\lambda''(x) = 0$ for $x \in [a', b']$. We have, however, by (81) and (68),

$$I_3'' = -\int_{-\infty}^{\infty} e^{ixu} d\Psi_2(u), \qquad \Psi_2(u) = \int_0^u t^2\, d\Psi(t), \tag{98}$$

$$I_1'' = -\int_{-\infty}^{\infty} e^{ixu} d\Phi_2(u) = \int_{-\infty}^{\infty} e^{ixu} d\varphi(u), \qquad \Phi_2(u) = \int_0^u t^2\, d\Phi(t). \tag{99}$$

By (88), (92), (95), (96), (83), (81), (98), (99), and (72), it follows that

$$-\int_{-\omega}^{\omega} e^{ixu} d\Psi_2(u) - \int_{-\omega}^{\omega} e^{ixu} d\psi(u) \to 0, \tag{100}$$

uniformly for $a' \leqslant x \leqslant b'$. By Theorem 4, however,

$$\int_{-\omega}^{\omega} e^{ixu} d\psi(u) - \int_{-\omega}^{\omega} e^{ixu} d\varphi(u) \to 0,$$

uniformly for $a' \leqslant x \leqslant b'$, since $\lambda(x)=1$ for $x \in [a', b']$. Hence (100) implies that

$$-\int_{-\omega}^{\omega} e^{ixu}\, d\Psi_2(u) - \int_{-\omega}^{\omega} e^{ixu}\, d\varphi(u) \to 0, \tag{101}$$

uniformly for $a' \leqslant x \leqslant b'$. The second integral in (101) occurs on the left-hand side of (91), so we have only to compute the first integral.

By Theorem 4, we have

$$\int_{-\omega}^{\omega} e^{ixu}\, d\,\Psi(u) - \lambda(x) \int_{-\omega}^{\omega} e^{ixu}\, d\Phi(u) \to 0\,,$$

as $\omega \to \infty$, uniformly in x. Hence, by (90),

$$\int_{-\infty}^{\infty} e^{ixt}\, \Psi'(t)\, dt = \int_{-\infty}^{\infty} e^{ixu}\, d\,\Psi(u) = F(x)\lambda(x)\,.$$

Since $\lambda(x)=0$ for $x \notin (a,b)$, and $\Psi \in C^{\infty}(-\infty, \infty)$ by (74), it follows that

$$\Psi'(u) = \frac{1}{2\pi} \int_{-\infty}^{\infty} e^{-itu} F(t)\lambda(t)\, dt = \frac{1}{2\pi} \int_{a}^{b} e^{-itu} F(t)\lambda(t)\, dt. \tag{102}$$

Hence

$$\begin{aligned} \int_{-\omega}^{\omega} e^{ixu}\, d\Psi(u) &= \frac{1}{2\pi} \int_{-\omega}^{\omega} e^{ixu}\, du \left(\int_{a}^{b} e^{-itu} F(t)\lambda(t)\, dt \right) \\ &= \frac{1}{2\pi} \int_{a}^{b} F(t)\lambda(t) \left(\int_{-\omega}^{\omega} e^{i(x-t)u}\, du \right) dt \\ &= \frac{1}{\pi} \int_{a}^{b} F(t)\lambda(t) \frac{\sin \omega(x-t)}{x-t}\, dt. \end{aligned} \tag{103}$$

Thus

$$-\int_{-\omega}^{\omega} e^{ixu}\, d\Psi_2(u) = \frac{1}{\pi} \int_{a}^{b} F(t)\lambda(t) \frac{d^2}{dt^2} \left(\frac{\sin \omega(x-t)}{x-t} \right) dt. \tag{104}$$

By (101) and (104) we have Theorem 5.

For the proof of the next theorem, we need the following

LEMMA. *If* (i) $f(x)$ *is of bounded variation in every finite interval in* $-\infty<x<\infty$, *and* $\beta_f(w)=o(w^{-2})$, *as* $|w|\to\infty$, *where*

$$\beta_f(w)=\int_w^{w+1} |df(x)|\,,$$

and (ii) $g(x)\in C^\infty(-\infty,\infty)$, *and*

$$g^{(r)}(x)=O\left(\frac{1}{1+|x|^k}\right), \tag{105}$$

for every $k>0$, *and* $r\geqslant 0$, *then*

$$\int_{-\infty}^{\infty} f(x-y)\,dg(y)=o(|x|^{-2}), \quad \textit{as} \quad |x|\to\infty\,. \tag{106}$$

If $g(x)=g(x,t)$, *where* t *is real, and* (105) *holds uniformly in* t, *then* (106) *also holds uniformly in* t.

PROOF. We suppose that $x\to+\infty$, for the proof in case $x\to-\infty$ is similar.

We split the integral in (106) into two parts, namely

$$\left(\int_{|y|\leqslant\frac{1}{2}x} + \int_{|y|>\frac{1}{2}x}\right) f(x-y)\,dg(y)\,, \qquad x>2\,. \tag{107}$$

The second integral, in absolute value, is less than

$$4x^{-2}\int_{|y|>\frac{1}{2}x} y^2|f(x-y)|\cdot|dg(y)|=O(x^{-2})\int_{|y|>\frac{1}{2}x} y^2|dg(y)|=o(x^{-2}), \tag{108}$$

as $x\to\infty$, since $f(y)=O(1)$ by hypothesis (i).

The first integral in (107) equals, on partial integration,

$$\begin{aligned}\{g(\tfrac{1}{2}x)f(\tfrac{1}{2}x)-g(-\tfrac{1}{2}x)f(\tfrac{3}{2}x)\} &- \int_{-\frac{1}{2}x}^{\frac{1}{2}x} g(y)\,d_y f(x-y)\\ =o(x^{-2}) &- \int_{-\frac{1}{2}x}^{\frac{1}{2}x} g(y)\,d_y f(x-y), \qquad \text{(by (105)}.\end{aligned} \tag{109}$$

Now

$$\begin{aligned}\left|\int_{-\frac{1}{2}x}^{\frac{1}{2}x} g(y)\,d_y f(x-y)\right| &\leqslant \sum_{r=0}^{m}\int_{-\frac{1}{2}x+r}^{-\frac{1}{2}x+r+1} |g(y)|\cdot|d_y f(x-y)|, \qquad m=[x]\\ &\leqslant \sum_{r=0}^{m} \max_{-\frac{1}{2}x+r\leqslant y\leqslant -\frac{1}{2}x+r+1} |g(y)|\cdot \int_{-\frac{1}{2}x+r}^{-\frac{1}{2}x+r+1} |d_y f(x-y)|\\ &= o(x^{-2}),\end{aligned} \tag{110}$$

since $x-y\geqslant\frac{1}{2}x-1$ in the last integral, and $\beta_f(w)=o(w^{-2})$ by hypothesis, and $\sum_{r=0}^{m}\max|g(y)|=O(1)$. From (110) and (109) it follows that the first integral in (107) is $o(x^{-2})$. This combined with (108) gives (106). The further conclusion is obvious from the uniformity in (105) and the method of proof of (106).

THEOREM 6 (ZYGMUND). *Let (a,b) be an interval contained in the interval $(0,2\pi)$. Let the functions φ, Φ, λ, and F be as in Theorem 5, with $\lambda(x)=1$ for $a<a'\leqslant x\leqslant b'<b$. Let $G(x)=F(x)\lambda(x)$ for $0<x\leqslant 2\pi$, and $G(x+2\pi)=G(x)$ for all $x\in(-\infty,\infty)$. Let the Fourier series of $G(x)$ be denoted by $c_0+\sum_{n=-\infty}^{\infty}{}'(in)^{-2}c_n e^{inx}$ (the dash on the sign of summation indicating that the term corresponding to $n=0$ is omitted). Then*

$$\sum_{-\omega\leqslant n\leqslant\omega}{}' c_n e^{inx}-\int_{-\omega}^{\omega} e^{ixu}\,d\varphi(u)\to 0, \tag{111}$$

as $\omega\to\infty$, uniformly for $a'\leqslant x\leqslant b'$. Further

$$c_n\to 0, \quad as \quad |n|\to\infty.$$

PROOF. If Ψ is defined as (73), then by (74) and (102), we have

$$\Psi'(u)=\frac{1}{2\pi}\int_0^{2\pi} e^{-itu}F(t)\lambda(t)\,dt \tag{112}$$

$$=\int_{-\infty}^{\infty}\mu'(u-y)\Phi(y)\,dy, \tag{113}$$

where the dash denotes ordinary differentiation.

Applying the above lemma with μ' in place of g, and Φ in place of f (which is allowed by (71) and (69)), we deduce that

$$\Psi'(u)=o(u^{-2}), \quad \text{as} \quad |u|\to\infty. \tag{114}$$

But $\Psi'(n)=c_n(in)^{-2}$, $n\neq 0$. Hence $c_n\to 0$ as $|n|\to\infty$. Clearly

$$\sum_{n=-N,n\neq 0}^{N} c_n e^{inx}=\frac{1}{\pi}\int_0^{2\pi} G(t)\cdot\frac{d^2}{dt^2}\left(\frac{\sin\{(N+\frac{1}{2})(x-t)\}}{2\sin\frac{1}{2}(x-t)}\right)dt. \tag{115}$$

By Theorem 5, with $\omega=N+\frac{1}{2}$,

$$\int_{-(N+\frac{1}{2})}^{N+\frac{1}{2}} e^{iux}\,d\varphi(u)-\frac{1}{\pi}\int_a^b F(t)\lambda(t)\frac{d^2}{dt^2}\left(\frac{\sin\{(N+\frac{1}{2})(x-t)\}}{x-t}\right)dt\to 0, \tag{116}$$

as $N\to\infty$, uniformly for $a'\leqslant x\leqslant b'$.

From (115) and (116) it is clear that in order to prove Theorem 6, we have only to show that

$$\int_a^b F(t)\lambda(t)\frac{d^2}{dt^2}\left[\left\{\frac{1}{2\sin\frac{1}{2}(x-t)}-\frac{1}{x-t}\right\}\sin\{(N+\tfrac{1}{2})(x-t)\}\right]dt\to 0, \quad (117)$$

as $N\to\infty$, uniformly for $a'\leqslant x\leqslant b'$.

Let us write

$$\Delta(u)=\frac{1}{2\sin(\frac{1}{2}u)}-\frac{1}{u}.$$

If we carry out the differentiation under the integral sign in (117), we obtain a sum of three terms, one of which is

$$-(N+\tfrac{1}{2})^2\int_a^b F(t)\lambda(t)\Delta(x-t)\sin\{(N+\tfrac{1}{2})(x-t)\}\,dt. \quad (118)$$

The other two terms contain lower powers of $(N+\frac{1}{2})$ and higher derivatives of Δ. We shall show that (118) tends to zero, uniformly for $a'\leqslant x\leqslant b'$, since the other two terms can be dealt with more easily.

If $x\in[a',b']$, and $t\in(a,b)$, then $x-t\in(-2\pi+\varepsilon,\ 2\pi-\varepsilon)$, $\varepsilon>0$. The function $\Delta(u)$ is regular for $-2\pi+\varepsilon<u<2\pi-\varepsilon$. Let $\Delta^*(u)\in C^\infty(-\infty,\infty)$, with $\Delta^*(u)=\Delta(u)$, for $|u|\leqslant 2\pi-\varepsilon$, and $\Delta^*(u)=0$ for $|u|\geqslant 2\pi$. We may then replace $\Delta(x-t)$ by $\Delta^*(x-t)$ in (118), provided that $a'\leqslant x\leqslant b'$.

Let

$$\delta(v)=\frac{1}{2\pi}\int_{-\infty}^{\infty}\Delta^*(u)e^{-iuv}\,du, \quad (119)$$

so that $\delta^{(r)}(v)=o\left(\frac{1}{1+|v|^k}\right)$, for every $k>0$, and integral $r\geqslant 0$, as $|v|\to\infty$. It follows that

$$e^{-ixv}\delta(-v)=\frac{1}{2\pi}\int_{-\infty}^{\infty}\Delta^*(x-u)e^{-iuv}\,du. \quad (120)$$

From (112) and (114), we have

$$\Psi'(u)=\frac{1}{2\pi}\int_a^b F(t)\lambda(t)e^{-iut}\,dt=o(u^{-2}). \quad (121)$$

Let

$$\chi_x(u)=\int_{-\infty}^{\infty}\Psi'(u-t)\delta(-t)e^{-ixt}\,dt. \quad (122)$$

Applying the above lemma with $\Psi(t)$ in place of $f(t)$, and $\delta(-t)e^{-ixt}$ in place of $g(t)$, we infer that $\chi_x(u)=o(u^{-2})$, uniformly in $x\in[a',b']$. Since

$$\frac{1}{2\pi}\int_{-\infty}^{\infty} F(t)\lambda(t)\Delta^*(x-t)e^{-i\omega t}\,dt=\chi_x(\omega)=o(\omega^{-2}),$$

as $|\omega|\to\infty$, uniformly for $a'\leqslant x\leqslant b'$, it follows that the term in (118) tends to zero uniformly for $a'\leqslant x\leqslant b'$, whence (117) and the theorem.

Corollary 1. *Consider the series* $\sum_{n=-\infty}^{\infty}{}' a_n e^{i\alpha_n x}$, *where*

$$\cdots\alpha_{-r}<\alpha_{-r+1}<\cdots<\alpha_{-1}<0<\alpha_1<\alpha_2<\cdots,$$

and

$$\sum_{n<\alpha_r\leqslant n+1}|a_r|=o(1),\quad as\quad |n|\to\infty. \tag{123}$$

Suppose that J is a closed interval contained in the interior of an interval I of length 2π, say $0<x\leqslant 2\pi$. Suppose that $\lambda(x)$ is a C^∞-function in $(-\infty,\infty)$, with compact support in I, which equals 1 on J. The series $\sum_{n=-\infty}^{\infty}{}' \frac{a_n}{(i\alpha_n)^2}e^{i\alpha_n x}$ *converges uniformly, and defines a continuous function, say $F(x)$. Let $G(x)$ be a periodic function of period 2π which coincides with $F(x)\lambda(x)$ on I. Let* $c_0+\sum_{n=-\infty}^{\infty}{}' \frac{c_n}{(in)^2}e^{inx}$ *be the Fourier series of $G(x)$. Then*

$$\sum_{|\alpha_n|\leqslant N}{}' a_n e^{i\alpha_n x}-\sum_{|n|\leqslant N}{}' c_n e^{inx}$$

converges uniformly to zero on J as $N\to\infty$. Further $c_n\to 0$, as $|n|\to\infty$.

We have only to choose $\varphi(u)=\sum_{0<\alpha_n\leqslant u} a_n$, if $u\geqslant 0$, and $\varphi(u)=-\sum_{u<\alpha_n<0} a_n$ if $u<0$, and apply Theorem 6.

We denote by $S(f)$ the Fourier series of a periodic function, of period 2π, which coincides with f in I.

Corollary 2. *With the same notation as in Corollary 1, if we suppose that* $\sum_{n=-\infty}^{\infty}{}' |a_n|<\infty$, *then the Fourier series of $g(x)$, which is of period 2π, and equals* $\sum_{n=-\infty}^{\infty}{}' a_n e^{i\alpha_n x}$ *in I, converges uniformly on J.*

PROOF. If $\sum_{n=-\infty}^{\infty} |a_n| < \infty$, then condition (123) is satisfied. Let $S(g)$ be the Fourier series of g, and let $g_2(x) = \sum_{n=-\infty}^{\infty}{}' (i\alpha_n)^{-2} a_n e^{i\alpha_n x}$, $g_1(x) = g_2'(x)$, the dash here denoting ordinary differentiation.

By Corollary 1 we know that $S''(\lambda \cdot g_2)$ is uniformly equiconvergent with $\sum a_n e^{i\alpha_n x}$. We have, however (because $\lambda^{(r)}(0) = \lambda^{(r)}(2\pi) = 0$, $r = 0, 1, 2, \ldots$),

$$S''(\lambda \cdot g_2) = S[(\lambda \cdot g_2)''] = S(\lambda \cdot g) + 2S(\lambda' \cdot g_1) + S(\lambda'' \cdot g_2),$$

where the dash denotes differentiation, and the dot the product. Since $\lambda' \cdot g_1$ and $\lambda'' \cdot g_2$ are continuously differentiable, the Fourier series $S(\lambda' \cdot g_1)$ and $S(\lambda'' \cdot g_2)$ are uniformly absolutely convergent. Hence $S''(\lambda \cdot g_2)$ is uniformly equiconvergent with $S(\lambda \cdot g)$ which, by Theorem 4′, is uniformly equiconvergent on J with $\lambda \cdot S(g)$. Since $\lambda(x) = 1$ for $x \in J$, and $\sum_{n=-\infty}^{\infty} a_n e^{i\alpha_n x}$ converges uniformly, it follows that $S(g)$ is uniformly convergent on J.

COROLLARY 3. *If in addition to condition* (123), *we assume that* $\sum_{n=-\infty}^{\infty}{}' |a_n| \alpha_n^{-1} < \infty$, *and* $q(x)$ *is any* C^∞*-function in* I, *then the series* $q(x) \cdot \sum_{n=-\infty}^{\infty}{}' a_n e^{i\alpha_n x}$ *is uniformly equiconvergent in* J *with the differentiated series of the Fourier series of a periodic function, of period* 2π, *which equals* $\lambda(x) \sum_{n=-\infty}^{\infty}{}' a_n W_n(x)$ *in* I, *where* $W_n(x) = \int_{x_0}^{x} q(t) e^{i\alpha_n t}\, dt$, $x_0 \in J$.

PROOF. We shall use a dot to denote a formal or ordinary product as the case may be, and a dash to denote a formal or ordinary derivative as appropriate.

Since the conclusion of the corollary is concerned with the interval J, we shall, for convenience, denote by q a periodic function, of period 2π, which belongs to the class $C^\infty(-\infty, \infty)$, and which equals q in J.

By partial integration, $W_n(x) = O(|\alpha_n|^{-1})$, as $|n| \to \infty$, for $x \in I$. Hence the series $\sum_{n=-\infty}^{\infty}{}' a_n W_n(x)$ converges, under the assumptions made on a_n.

The periodic function given in $0 < x \leqslant 2\pi$ by

$$G(x) = \lambda(x) \cdot \sum_{n=-\infty}^{\infty}{}' \frac{a_n}{(i\alpha_n)^2} e^{i\alpha_n x} = \lambda(x) \cdot F(x),$$

say, is continuously differentiable, and $S(q \cdot G) = S(q) \cdot S(G)$.

In order to prove that $q(x)\cdot \sum_{n=-\infty}^{\infty}{}' a_n e^{i\alpha_n x}$ and $S'(\lambda(x)\cdot \sum_{n=-\infty}^{\infty}{}' a_n W_n(x))$ are uniformly equiconvergent for $x\in J$, we shall first prove that

$$S''(q\cdot G) \quad \textit{and} \quad S(q)\cdot S''(G) \quad \textit{are uniformly equiconvergent on } J; \tag{124}$$

secondly, that

$$S''(q\cdot G) \quad \textit{and} \quad S''\left(\lambda\cdot \sum_{n=-\infty}^{\infty}{}' a_n v_n\right) \quad \textit{are uniformly equiconvergent on } J, \tag{125}$$

where

$$v_n(x)=\int_{x_0}^{x} W_n(t)\,dt, \qquad x_0\in J;$$

and thirdly, that

$$S'\left(\lambda\cdot \sum_{n=-\infty}^{\infty}{}' a_n v_n\right) \quad \textit{and} \quad S'\left(\lambda\cdot \sum_{n=-\infty}^{\infty}{}' a_n W_n\right) \quad \textit{are uniformly equiconvergent on } J. \tag{126}$$

From (124)—(126), it would follow that

$$S(q)\cdot S''(G) \quad \textit{and} \quad S'\left(\lambda\cdot \sum_{n=-\infty}^{\infty}{}' a_n W_n\right) \quad \textit{are uniformly equiconvergent on } J. \tag{127}$$

But $G=F\cdot\lambda$, and the coefficients of the series $S''(G)$ tend to zero as $|n\to\infty$, because of Corollary 1. By Theorem 4′ and Corollary 1, it follows that the product $S(q)\cdot S''(G)$ is uniformly equiconvergent on J with $q(x)\cdot \sum_{n=-\infty}^{\infty}{}' a_n e^{i\alpha_n x}$. Hence, by (127), it would follow that $q(x)\cdot \sum_{n=-\infty}^{\infty}{}' a_n e^{i\alpha_n x}$ and $S'(\lambda(x)\cdot \sum_{n=-\infty}^{\infty}{}' a_n W_n(x))$ are uniformly equiconvergent on J.

To complete the proof of the corollary, we have now only to prove (124)—(126).

To prove (124), we note that, by (89),

$$S''(q\cdot G)=(S(q)\cdot S(G))''=S''(q)\cdot S(G)+2S'(q)\cdot S'(G)+S(q)\cdot S''(G).$$

Since $G=F\cdot\lambda$, $S'(q)=S(q')$, and $S''(q)=S(q'')$, we have

$$S''(q\cdot G)=S(q'')\cdot S(G)+2S(q')\cdot S(F')\cdot S(\lambda)+2S(q')\cdot S(F)\cdot S(\lambda')+S(q)\cdot S''(G).$$

Since G is continuously differentiable, and q is a C^∞-function, $S(q'')\cdot S(G)$ $=S(q''\cdot G)$, which converges uniformly in J. And $S(q')\cdot S(F)\cdot S(\lambda')$ $=S(q'\cdot F\cdot\lambda')$, which converges uniformly. Further $\lambda\cdot q'\in C^\infty(-\infty,\infty)$, so that Theorem 4′ is applicable to the product $S(\lambda\cdot q')\cdot S(F')$, which is uniformly equiconvergent with $\lambda\cdot q'\cdot S(F')$, while $S(F')$ converges uniformly by Corollary 2. Assertion (124) follows.

To prove (125), we take the general term of the series $(q\cdot G) - \left(\lambda\cdot \sum_{n=-\infty}^{\infty}{}' a_n v_n\right)$, without the factor a_n, and note that

$$\left\{\lambda(x)\left(\frac{q(x)e^{i\alpha_n x}}{(i\alpha_n)^2} - v_n(x)\right)\right\}''$$

$$=(\lambda(x)\cdot q(x))''\frac{e^{i\alpha_n x}}{(i\alpha_n)^2} - \lambda''(x)\cdot v_n(x)+2\lambda'(x)\cdot\frac{q(x_0)e^{i\alpha_n x_0}}{(i\alpha_n)}+$$

$$+\frac{2\lambda'(x)}{i\alpha_n}\int_{x_0}^{x} q'(t)e^{i\alpha_n t}dt+2\lambda(x)q'(x)\frac{e^{i\alpha_n x}}{(i\alpha_n)}.$$

Here the sum-function of the series arising from the first term, namely $\sum_{n=-\infty}^{\infty}{}' (i\alpha_n)^{-2}a_n e^{i\alpha_n x}$, is continuously differentiable. So is the sum-function of the series $\sum_{n=-\infty}^{\infty}{}' a_n v_n(x)$, since $v_n'(x)=O(|\alpha_n|^{-1})$, and $\sum_{n=-\infty}^{\infty}{}' a_n v_n'(x)$ is absolutely, uniformly convergent. The series $\sum_{n=-\infty}^{\infty}{}' (i\alpha_n)^{-1}a_n e^{i\alpha_n x_0}$ absolutely converges to a constant. The sum-function of the series $\sum_{n=-\infty}^{\infty}{}' a_n(i\alpha_n)^{-1}\int_{x_0}^{x} q'(t)e^{i\alpha_n t}dt$ again is continuously differentiable. By Corollary 2, and Theorem 4′, on the other hand, $S\left(\lambda(x)\cdot q'(x)\cdot \sum_{n=-\infty}^{\infty}{}' (i\alpha_n)^{-1}a_n e^{i\alpha_n x}\right)$ converges uniformly for $x\in J$. Hence the difference $S''(q\cdot G)-S''\left(\lambda\cdot \sum_{n=-\infty}^{\infty}{}' a_n v_n\right)$ converges uniformly on J, and (125) is proved. The proof of (126) is similar, and the corollary follows.

Remark. In this section, we have taken I to be the interval $0<x\leqslant 2\pi$, and J to be a closed interval contained in the interior of I. But the results are valid with greater generality. For, given an interval $J':-\infty<y_1\leqslant y\leqslant y_2<+\infty$, contained in the interior of an interval I', we can, by a linear transformation, map I' into the interval $I:0<x\leqslant 2\pi$. We shall make use of this fact in the next section.

§ 9. The Voronoi identity. In Theorem 1 we proved the identity

$$D^\alpha(y^2)-Q_\alpha(y^2)=\pi^{-1-2\alpha}\sum_{n=1}^{\infty} d(n)n^{-1-\alpha}f_\alpha(\pi^2 n y^2),\qquad y>0,\quad (128)$$

where Q_α is a C^∞-function, for $\alpha > \frac{1}{2}$. We shall now consider the validity of this for $\alpha > -\frac{1}{2}$. Specifically we shall consider the convergence of the series

$$\sum_{n=1}^{\infty} d(n) n^{-\nu} f_{\nu-1}(\pi^2 n x), \qquad x > 0, \tag{129}$$

under the two conditions

$$\sum_{n=1}^{\infty} \frac{d(n)}{n^{\frac{1}{2}\nu + \frac{3}{4}}} < \infty, \tag{130}$$

and

$$\sum_{m^2 < n \leqslant (m+1)^2} d(n) \cdot n^{-\frac{1}{2}\nu - \frac{1}{4}} = o(1), \quad \text{as} \quad m \to \infty. \tag{131}$$

These conditions are satisfied if, and only if, $\nu > \frac{1}{2}$.

By formula (30), we have, for $\rho \geqslant -\frac{3}{2}$, $x > 0$,

$$f_\rho(x) = \kappa_0 x^{\frac{1}{2}\rho + \frac{1}{4}} \cos(4x^{\frac{1}{2}} - \tfrac{1}{2}\rho\pi - \tfrac{1}{4}\pi) + \kappa_1 x^{\frac{1}{2}\rho - \frac{1}{4}} \cos(4x^{\frac{1}{2}} - \tfrac{1}{2}\rho\pi + \tfrac{1}{4}\pi) + O(x^{\frac{1}{2}\rho - \frac{3}{4}}). \tag{132}$$

If we note that the constant c_1 in (24)′ is given by $c_1 = A_0 + \frac{1}{2}\alpha(\alpha - 1)$, where A_0 is an absolute constant, it follows that $\kappa_0 = (2\pi)^{-\frac{1}{2}} 2^{-\rho}$, $\kappa_1 = (2\pi)^{-\frac{1}{2}} 2^{-\rho-1}(A_0 + \frac{1}{4}\rho(\rho+2))$. Hence

$$\begin{aligned} f_{\nu-1}(\pi^2 n y^2) = {} & c_1 y^{\nu - \frac{1}{2}} n^{\frac{1}{2}\nu - \frac{1}{4}} \cos(4\pi y n^{\frac{1}{2}} - \tfrac{1}{2}\nu\pi + \tfrac{1}{4}\pi) + \\ & + c_2 y^{\nu - \frac{3}{2}} n^{\frac{1}{2}\nu - \frac{3}{4}} \cos(4\pi y n^{\frac{1}{2}} - \tfrac{1}{2}\nu\pi + \tfrac{3}{4}\pi) + O(n^{\frac{1}{2}\nu - \frac{5}{4}}), \end{aligned} \tag{133}$$

where $c_1 = \pi^{\nu-1} 2^{-\nu+\frac{1}{2}}$, $c_2 = \pi^{\nu-2} 2^{-\nu-\frac{1}{2}} (A_0 + \frac{1}{4}(\nu-1)(\nu+1))$, uniformly in any bounded interval in $y > 0$.

The function P defined by

$$\begin{aligned} P(y) = 2y \cdot \sum_{n=1}^{\infty} d(n) n^{-\nu} \{ & f_{\nu-1}(\pi^2 n y^2) - \\ & - c_1 y^{\nu-\frac{1}{2}} n^{\frac{1}{2}\nu - \frac{1}{4}} \cos(4\pi y n^{\frac{1}{2}} - \tfrac{1}{2}\nu\pi + \tfrac{1}{4}\pi) - \\ & - c_2 y^{\nu - \frac{3}{2}} n^{\frac{1}{2}\nu - \frac{3}{4}} \cos(4\pi y n^{\frac{1}{2}} - \tfrac{1}{2}\nu\pi + \tfrac{3}{4}\pi)\} \end{aligned} \tag{134}$$

is a continuously differentiable function of y for $y > 0$, because of (20), and (133) with $\nu - 1$ in place of ν.

By an application of Corollary 3 of Theorem 6 (see the Remark at the end of § 8), with $\alpha_n = 4\pi n^{\frac{1}{2}}$, $\alpha_{-n} = -\alpha_n$, the series

$$2y \sum_{n=1}^{\infty} d(n) y^{\nu - \frac{1}{2}} n^{-\frac{1}{2}\nu - \frac{1}{4}} \cos(4\pi y n^{\frac{1}{2}} - \tfrac{1}{2}\nu\pi + \tfrac{1}{4}\pi)$$

is uniformly equiconvergent on J with the derived series of the Fourier series of a periodic function which in I equals

$$\lambda(y)\cdot\sum_{n=1}^{\infty} d(n)\int_{y_0}^{y} 2\xi\cdot\xi^{\nu-\frac{1}{2}} n^{-\frac{1}{2}\nu-\frac{1}{4}}\cos(4\pi\xi n^{\frac{1}{2}}-\tfrac{1}{2}\nu\pi+\tfrac{1}{4}\pi)\,d\xi,$$

where $y_0 \in J$. The corresponding statement for the last series on the right-hand side of (134) is similarly valid.

Hence the series

$$2y\cdot\sum_{n=1}^{\infty} d(n)n^{-\nu} f_{\nu-1}(\pi^2 n y^2)$$

is uniformly equiconvergent on J with the derived series of the Fourier series of a periodic function which in I equals

$$\begin{aligned}&\lambda(y)\sum_{n=1}^{\infty} d(n)n^{-\nu}\int_{y_0}^{y} 2\eta f_{\nu-1}(\pi^2 n\eta^2)\,d\eta \\ &=\lambda(y)\pi^{-2}\sum_{n=1}^{\infty} d(n)n^{-\nu-1}(f_\nu(\pi^2 n y^2)-f_\nu(\pi^2 n y_0^2))=\lambda(y)\pi^{-2}(S_{\nu+1}(y)-K),\end{aligned} \tag{135}$$

where K is a constant depending on y_0, and

$$S_\nu(y)\equiv\sum_{n=1}^{\infty} d(n)n^{-\nu} f_{\nu-1}(\pi^2 n y^2). \tag{136}$$

Thus we have the following

THEOREM 7. *If* $\nu>\frac{1}{2}$, *the series* $2\pi^2 y S_\nu(y)$ *is uniformly equiconvergent on* J *with the derived series of the Fourier series of a periodic function which in* I *equals* $\lambda(y)(S_{\nu+1}(y)-K)$, *and the function* $S_{\nu+1}(y)$ *is continuous for* $y>0$.

We shall use this result to study the identity

$$D^\alpha(x)-Q_\alpha(x)=\pi\sum_{n=1}^{\infty} d(n)(\pi^2 n)^{-1-\alpha} f_\alpha(\pi^2 n x),\quad x>0, \tag{32}$$

for $\alpha>-\frac{1}{2}$.

THEOREM 8. *The series on the right-hand side of* (32) *converges for every* $\alpha>-\frac{1}{2}$, *uniformly in any closed interval in* $x>0$ *in which the function on the left-hand side is continuous, and boundedly in any interval* $0<x_1\leqslant x\leqslant x_2<\infty$ *if* $\alpha=0$.

PROOF. By Theorem 7, the series $2\pi^2 y S_{\alpha+1}(y)$ is uniformly equiconvergent on J with the derived series of the Fourier series of a periodic function which in I equals $\lambda(y)(S_{\alpha+2}(y)-K)$, provided that $\alpha+1>\frac{1}{2}$. By Theorem 1 and (136), we have, for $\alpha+1>\frac{1}{2}$,

$$\lambda(y)\pi^{-2\alpha-3}S_{\alpha+2}(y)$$

$$=\frac{\lambda(y)}{\Gamma(\alpha+2)}\sum_{n<y^2} d(n)(y^2-n)^{\alpha+1}-Q_{\alpha+1}(y^2)\lambda(y)$$

$$=\frac{\lambda(y)}{\Gamma(\alpha+1)}\int_{y_1}^{y^2}\sum_{n<\eta} d(n)(\eta-n)^{\alpha}\,d\eta-Q_{\alpha+1}(y^2)\lambda(y),\qquad 0<y_1<1$$

$$=\frac{\lambda(y)}{\Gamma(\alpha+1)}\int_{y_2}^{y}\sum_{n<\eta^2} d(n)(\eta^2-n)^{\alpha}\cdot 2\eta\,d\eta-Q_{\alpha+1}(y^2)\lambda(y),\quad 0<y_2<1.$$

The theorem now follows from the properties of the Fourier series of a periodic function which equals $\frac{\lambda(\eta)\cdot 2\eta}{\Gamma(\alpha+1)}\sum_{n<\eta^2} d(n)(\eta^2-n)^{\alpha}$ in I, and from the fact that $\lambda(y)=1$ for $y\in J$, since $Q_{\alpha+1}(y^2)\in C^{\infty}(0<y<\infty)$.

REMARKS. (i) As a consequence of Theorem 8, identity (32) holds for $\alpha>-\frac{1}{2}$, if for $\alpha\leqslant 0$ integers x are excluded. This result, for $\alpha=0$, was proved first by Voronoi and later by Hardy.

(2) For the proof of Theorem 8 we have used, apart from the theorem of Zygmund (Theorem 6), the functional equation of $\zeta^2(s)$, and the elementary result that $d(n)=O(n^{\varepsilon})$ for every $\varepsilon>0$.

Notes on Chapter VIII

As general references see G. Voronoi, *Annales de l'École Norm.* (3) 21 (1904), 207—268; 459—534; G. H. Hardy, *Proc. London Math. Soc.* (2) 15 (1916), 1—25; 192—213; E. Landau, *Vorlesungen*, II, 240—249; A. E. Ingham, *Proc. Cambridge Phil. Soc.* 36 (1940), 131—138; K. Chandrasekharan and Raghavan Narasimhan, *Annals of Math.* 74 (1961), 1—23; 76 (1962), 93—136; *Math. Annalen*, 152 (1963), 30—64. The exposition of this chapter is based on the papers by the author and Narasimhan, and on several subsequent discussions.

§ 1. For fuller references on the best-known value of θ, see the article by Bohr and Cramér in *Enzykl. d. Math. Wiss.* IIc 8 (1922), 823—824, as well as the article by L. K. Hua, *Enzykl. d. Math. Wiss.* Bd. 12, Heft 13, Teil 1. Dirichlet proved in 1849 that $\theta \leqslant \frac{1}{2}$; Voronoi in 1904 that $\theta \leqslant \frac{1}{3}$; Hardy and Landau in 1916 that $\theta \geqslant \frac{1}{4}$; van der Corput in 1922 that $\theta < \frac{33}{100}$, and in 1928 that $\theta \leqslant \frac{27}{82}$; T. T. Chih (1950) and H. E. Richert (1953) that $\theta \leqslant \frac{15}{46}$. For elementary facts about the divisor function, see, for instance, the author's *Introduction*, Ch. VI.

§§ 2—3. See Chandrasekharan and Narasimhan, *Annals of Math.* 76 (1962), 93—136. Perron's formula, in the form used, is given, for instance, in *Typical means* (Oxford, 1952) by Chandrasekharan and Minakshisundaram, Lemma 3.65.

Formula (17) is a special case of a general formula stated and used by Chandrasekharan and Narasimhan, *Math. Annalen*, 152 (1963), 30—64. Formula (30) is a special case of Lemma 1 there.

§ 4. Formula (39) is proved by the author and Narasimhan, *Annals of Math.* 76 (1962), Lemma 7.1.

§ 5. See Voronoi, loc. cit. The proof given here is a special case of Theorem 4.1 in the paper by Narasimhan and the author, *Annals of Math.* 76 (1962), Theorem 4.1.

§ 6. See A. S. Besicovitch, *J. London Math. Soc.* 15 (1940), 3—6. The proof of the special case treated here was communicated to the author by Prof. Narasimhan. A *square-free* integer is one which is not divisible by the square of a prime.

§ 7. See Hardy, Landau, and Ingham, cited above. The proof of Theorem 3 given here is modelled on the more general one given in the *Annals of Math.* 76 (1962), Theorems 3.1 and 3.2, by the author and Narasimhan. For further work on Theorem 3, see K. S. Gangadharan, *Proc. Cambridge Phil. Soc.* 57 (1961), 699—721; K. A. Corrádi and I. Kátai, *Magyar Tud. Akad. Mat. Fiz. Oszt. Közl.* 17 (1967), 89—97.

§ 8. For the equiconvergence theorems treated here, see A. Zygmund, *Annals of Math.* 48 (1947), 393—440; as well as his book *Trigonometric series* (Cambridge) 1959, I, 330—337; II, 286—290. Corollaries 1, 2, 3, which are crucial for the application in §9, are modelled on the corresponding results in the *Annals of Math.* 74 (1961), 1—23, by the author and Narasimhan. For Theorem 4′, see A. Rajchman, *Comptes Rendus, Varsovie*, 11 (1918), 115—122; *Math. Annalen*, 95 (1926), 388 – 408.

§ 9. The proofs of Theorems 7 and 8 are parallel to the proofs of Theorems II and III, *Annals of Math.* 74 (1961), 1—23, due to the author and Narasimhan. They have *used* the corresponding identity, with $d(n)$ replaced by the ideal function in a real quadratic field, to prove a summation formula, *Commentarii Math. Helvetici*, 43 (1968), 296—310, Theorem 2.

The constants κ_0, κ_1 in (132) have been computed by H. Joris (in a paper to appear) by reckoning the constants in Stirling's formula (24)'.

A list of books

[1] Bohr, H., and Cramér, H.: *Die neuere Entwicklung der analytischen Zahlentheorie*. Enzyklopädie der mathematischen Wissenschaften, II. 3, Heft 6. Leipzig: Teubner 1923.

[2] Chandrasekharan, K.: *Introduction to analytic number theory*. Berlin-Heidelberg-New York: Springer 1968 (referred to as the *Introduction*).

[3] Davenport, H.: *Multiplicative number theory*. Chicago: Markham 1967.

[4] Dickson, L. E.: *History of the theory of numbers*. (Carnegie Institution, Washington) I (1919), II (1920), III (1923), reprinted New York: Chelsea 1952.

[5] Estermann, T.: *Introduction to modern prime number theory*. Cambridge Tracts, No. 41, Cambridge 1940.

[6] Hardy, G. H.: *Ramanujan*. Cambridge 1940.

[7] Hardy, G. H.: *Collected papers* (including joint papers with J. E. Littlewood and others). I, II, Oxford 1966—67.

[8] Hecke, E.: *Mathematische Werke*. Göttingen 1959.

[9] Hua, L. K.: *Die Abschätzung von Exponentialsummen und ihre Anwendung in der Zahlentheorie*. Enzyklopädie der mathematischen Wissenschaften, I. 2, Heft 13, Teil 1, Leipzig: Teubner 1959.

[10] Hua, L. K.: *Additive Primzahltheorie*. Leipzig 1959.

[11] Ingham, A. E.: *The distribution of prime numbers*. Cambridge Tracts, No. 30. Cambridge 1932 (referred to as Ingham's *Tract*).

[12] Landau, E.: *Handbuch der Lehre von der Verteilung der Primzahlen*, Second Edition. New York: Chelsea 1953 (referred to as Landau's *Primzahlen*).

[13] Landau, E.: *Vorlesungen über Zahlentheorie*, I, II, III. Leipzig: Hirzel 1927 (referred to as Landau's *Vorlesungen*).

[14] Landau, E.: *Über einige neuere Fortschritte der additiven Zahlentheorie*. Cambridge Tracts, No. 35, 1937, reprinted (Stechert-Hafner, New York 1964).

[15] Littlewood, J. E.: *Lectures on the theory of functions*. Oxford 1944.

[16] Prachar, K.: *Primzahlverteilung*. Berlin-Göttingen-Heidelberg: Springer 1957.

[17] Ramanujan, S.: *Collected papers*. Cambridge 1927.

[18] Riemann, B.: *Gesammelte mathematische Werke*. 1876.

[19] Siegel, C. L.: *Gesammelte Abhandlungen*. Berlin-Heidelberg-New York: Springer 1966.

[20] Titchmarsh, E. C.: *The theory of the Riemann Zeta-function*. Oxford 1951. (referred to as Titchmarsh's *Zeta-function*).

[21] Turán, P.: *Eine neue Methode in der Analysis und deren Anwendungen*. Budapest 1953.

[22] Vinogradov, I. M.: *The method of trigonometrical sums in the theory of numbers*, English translation. New York: Interscience 1955.

[23] Walfisz, A.: *Weylsche Exponentialsummen in der neueren Zahlentheorie*. Berlin 1963.

[24] Weyl, H.: *Gesammelte Abhandlungen*. Berlin-Heidelberg-New York: Springer 1968.

[25] Zygmund, A.: *Trigonometric series*, I, II. Cambridge 1959.

Subject Index

Abel's summation formula 22
arithmetical function 1

Bertrand's postulate 92
Besicovitch's theorem 204
Bessel functions 195
Borel-Carathéodory, Lemma of
  41

character, conductor of 147
—, derived 146
—, equivalent 147
—, extended 146
—, improper 147
—, principal 143
—, proper 147
Chebyshev's functions ϑ, ψ 3
Chudakov's theorem 110

Dedekind's η-function 174
Dirichlet's divisor problem 194
— L-functions 144
— series 23
— theorem 145
— divisor function $d(n)$ 19
— estimate of $d(n)$ 7, 24

entire function, order of 40
error term 27, 59
Euler's function φ 25
— identity 28

Farey sequence 185
Fejér's theorem 29
functional equation 28, 149

Hadamard, factorization
  theorem of 42
Hamburger's theorem 51
Hardy-Ingham-Pólya, a theorem
  of 126
Hardy's theorem 48
Hardy-Ramanujan formula 166
Hecke's series 155
Hilbert's inequality 26
Hoheisel's theorem 124

Ingham's theorem 134

Kronecker's symbol 163

Landau's lemma 75
— theorem 78
— formula for ψ 120
Legendre symbol 163
Lindelöf hypothesis 113
Littlewood's theorem 77, 140
logarithmic integral li 59

von Mangoldt's function Λ 3
maximum modulus principle 35
Möbius's function μ 1
Möbius inversion formula,
  first 2
— — —, second 2
— — —, a variant of the
  second 3
modular transformation 175

orthogonality relations 144

partition function $p(n)$ 166
— —, the generating function of the 168
Perron's formula 80, 87, 195
Phragmén-Lindelöf principle 53
Poisson's summation formula 28
prime number theorem 1
— — — for arithmetical progressions 25, 145

quadratic field, class-number of 163
— —, discriminant of 163
— —, fundamental unit of 163

Rademacher's identity 166
Rajchman's theorem 213
Riemann hypothesis 51
— von Mangoldt formula 36

Schwarz's inequality 12
Selberg's formula 4
— —, a variant of 6
Siegel's partial-fraction formula 57
— theorem 161
standard form 19
Stieltjes integral 7

theta-relation 29
trigonometric series,
— —, equiconvergence of 209
— —, formal derivative of 212
— —, formal product of 212
trivial zeros 33, 154

Vinogradov's inequality 100
— mean-value theorem 97
— theorem 164
Voronoi's theorem 202
— identity 223

Weyl's inequality 62
— sum 60
Wirsing's inequality 12

Zeta-function of Riemann 28
—, critical line of 33
—, critical strip of 33
—, functional equation of 28
—, trivial zeros of 33
Zygmund's theorem 214, 218

Die Grundlehren der mathematischen Wissenschaften in Einzeldarstellungen mit besonderer Berücksichtigung der Anwendungsgebiete

---

2. Knopp: Theorie und Anwendung der unendlichen Reihen. DM 48,—; US $ 13.20
3. Hurwitz: Vorlesungen über allgemeine Funktionentheorie und elliptische Funktionen. DM 49,—; US $ 13.50
4. Madelung: Die mathematischen Hilfsmittel des Physikers. DM 49,70; US $ 13.70
10. Schouten: Ricci-Calculus. DM 58,60; US $ 16.20
14. Klein: Elementarmathematik vom höheren Standpunkte aus. 1. Band: Arithmetik, Algebra, Analysis. DM 24,—; US $ 6.60
15. Klein: Elementarmathematik vom höheren Standpunkte aus. 2. Band: Geometrie. DM 24,—; US $ 6.60
16. Klein: Elementarmathematik vom höheren Standpunkte aus. 3. Band: Präzisions- und Approximationsmathematik. DM 19,80; US $ 5.50
20. Pólya/Szegö: Aufgaben und Lehrsätze aus der Analysis II: Funktionentheorie, Nullstellen, Polynome, Determinanten, Zahlentheorie. DM 38,—; US $ 10.50
22. Klein: Vorlesungen über höhere Geometrie. DM 28,—; US $ 7.70
26. Klein: Vorlesungen über nicht-euklidische Geometrie. DM 24,—; US $ 6.60
27. Hilbert/Ackermann: Grundzüge der theoretischen Logik. DM 38,—; US $ 10.50
30. Lichtenstein: Grundlagen der Hydromechanik. DM 38,—; US $ 10.50
31. Kellogg: Foundations of Potential Theory. DM 32,—; US $ 8.80
32. Reidemeister: Vorlesungen über Grundlagen der Geometrie. DM 18,—; US $ 5.00
38. Neumann: Mathematische Grundlagen der Quantenmechanik. DM 28,—; US $ 7.70
40. Hilbert/Bernays: Grundlagen der Mathematik I. DM 68,—; US $ 18.70
43. Neugebauer: Vorlesungen über Geschichte der antiken mathematischen Wissenschaften. 1. Band: Vorgriechische Mathematik. DM 48,—; US $ 13.20
50. Hilbert/Bernays: Grundlagen der Mathematik II. DM 68.—; US $ 18.70
52. Magnus/Oberhettinger/Soni: Formulas and Theorems for the Special Functions of Mathematical Physics. DM 66,—; US $ 16.50
57. Hamel: Theoretische Mechanik. DM 84,—; US $ 23.10
58. Blaschke/Reichardt: Einführung in die Differentialgeometrie. DM 24,—; US $ 6.60
59. Hasse: Vorlesungen über Zahlentheorie. DM 69,—; US $ 19.00
60. Collatz: The Numerical Treatment of Differential Equations. DM 78,—; US $ 19.50
61. Maak: Fastperiodische Funktionen. DM 38,—; US $ 10.50
62. Sauer: Anfangswertprobleme bei partiellen Differentialgleichungen. DM 41,—; US $ 11.30
64. Nevanlinna: Uniformisierung. DM 49,50; US $ 13.70
66. Bieberbach: Theorie der gewöhnlichen Differentialgleichungen. DM 58,50; US $ 16.20
68. Aumann: Reelle Funktionen. DM 68,—; US $ 18.70
69. Schmidt: Mathematische Gesetze der Logik I. DM 79,—; US $ 21.80
71. Meixner/Schäfke: Mathieusche Funktionen und Sphäroidfunktionen mit Anwendungen auf physikalische und technische Probleme. DM 52,60; US $ 14.50
73. Hermes: Einführung in die Verbandstheorie. DM 46,—; US $ 12.70
74. Boerner: Darstellungen von Gruppen. DM 58,—; US $ 16.00
75. Rado/Reichelderfer: Continuous Transformations in Analysis, with an Introduction to Algebraic Topology. DM 59,60; US $ 16.40
76. Tricomi: Vorlesungen über Orthogonalreihen. DM 68,—; US $ 18.70

77. Behnke/Sommer: Theorie der analytischen Funktionen einer komplexen Veränderlichen. DM 79,—; US $ 21.80
78. Lorenzen: Einführung in die operative Logik und Mathematik. DM 54,—; US $ 14.90
80. Pickert: Projektive Ebenen. DM 48,60; US $ 13.40
81. Schneider: Einführung in die transzendenten Zahlen. DM 24,80; US $ 6.90
82. Specht: Gruppentheorie. DM 69,60; US $ 19.20
84. Conforto: Abelsche Funktionen und algebraische Geometrie. DM 41,80; US $ 11.50
86. Richter: Wahrscheinlichkeitstheorie. DM 68,—; US $ 18,70
88. Müller: Grundprobleme der mathematischen Theorie elektromagnetischer Schwingungen. DM 52,80; US $ 14.60
89. Pfluger: Theorie der Riemannschen Flächen. DM 39,20; US $ 10.80
90. Oberhettinger: Tabellen zur Fourier-Transformation. DM 39,50; US $ 10.90
91. Prachar: Primzahlverteilung. DM 58,—; US $ 16.00
93. Hadwiger: Vorlesungen über Inhalt, Oberfläche und Isoperimetrie. DM 49,80; US $ 13.70
94. Funk: Variationsrechnung und ihre Anwendung in Physik und Technik. DM 120,—; US $ 33.00
95. Maeda: Kontinuierliche Geometrien. DM 39,—; US $ 10.80
97. Greub: Linear Algebra. DM 39,20; US $ 9.80
98. Saxer: Versicherungsmathematik. 2. Teil. DM 48,60; US $ 13.40
99. Cassels: An Introduction to the Geometry of Numbers. DM 69,—; US $ 19.00
100. Koppenfels/Stallmann: Praxis der konformen Abbildung. DM 69,—; US $ 19.00
101. Rund: The Differential Geometry of Finsler Spaces. DM 59,60; US $ 16.40
103. Schütte: Beweistheorie. DM 48,—; US $ 13.20
104. Chung: Markov Chains with Stationary Transition Probabilities. DM 56,—; US $ 14.00
105. Rinow: Die innere Geometrie der metrischen Räume. DM 83,—; US $ 22.90
106. Scholz/Hasenjaeger: Grundzüge der mathematischen Logik. DM 98,—; US $ 27.00
107. Köthe: Topologische lineare Räume I. DM 78,—; US $ 21.50
108. Dynkin: Die Grundlagen der Theorie der Markoffschen Prozesse. DM 33,80; US $ 9.30
110. Dinghas: Vorlesungen über Funktionentheorie. DM 69,—; US $ 19.00
112. Morgenstern/Szabó: Vorlesungen über theoretische Mechanik. DM 69,—; US $ 19.00
113. Meschkowski: Hilbertsche Räume mit Kernfunktion. DM 58,—; US $ 16.00
114. MacLane: Homology. DM 62,—; US $ 15.50
115. Hewitt/Ross: Abstract Harmonic Analysis. Vol. 1: Structure of Topological Groups, Integration Theory, Group Representations. DM 76,—; US $ 20.90
116. Hörmander: Linear Partial Differential Operators. DM 42,—; US $ 10.50
117. O'Meara: Introduction to Quadratic Forms. DM 48,—; US $ 13.20
118. Schäfke: Einführung in die Theorie der speziellen Funktionen der mathematischen Physik. DM 49,40; US $ 13.60
119. Harris: The Theory of Branching Processes. DM 36,—; US $ 9.90
120. Collatz: Funktionalanalysis und numerische Mathematik. DM 58, ; US $ 16.00
121.
122. Dynkin: Markov Processes. DM 96,—; US $ 26.40
123. Yosida: Functional Analysis. DM 66,—; US $ 16.50
124. Morgenstern: Einführung in die Wahrscheinlichkeitsrechnung und mathematische Statistik. DM 38,—; US $ 10.50
125. Itô/McKean: Diffusion Processes and Their Sample Paths. DM 58,—; US $ 16.00
126. Lehto/Virtanen: Quasikonforme Abbildungen. DM 38,—; US $ 10.50
127. Hermes: Enumerability, Decidability, Computability. DM 39,—; US $ 10.80

128. Braun/Koecher: Jordan-Algebren. DM 48,—; US $ 13.20
129. Nikodým: The Mathematical Apparatus for Quantum-Theories. DM 144,—; US $ 36.00
130. Morry: Multiple Integrals in the Calculus of Variations. DM 78,—; US $ 19.50
131. Hirzebruch: Topological Methods in Algebraic Geometry. DM 38,—; US $ 9.50
132. Kato: Perturbation Theory for Linear Operators. DM 79,20; US $ 19.80
133. Haupt/Künneth: Geometrische Ordnungen. DM 68,—; US $ 18.70
134. Huppert: Endliche Gruppen I. DM 156,—; US $ 42.90
135. Handbook for Automatic Computation. Vol 1/Part a: Rutishauser: Description of ALOGL 60. DM 58,—; US $ 14.50
136. Greub: Multilinear Algebra. DM 32,—; US $ 8.00
137. Handbook for Automatic Computation. Vol. 1/Part b: Grau/Hill/Langmaack: Translation of ALOGL 60. DM 64,—; US $ 16.00
138. Hahn: Stability of Motion. DM 72,—; US $ 19.80
139. Mathematische Hilfsmittel des Ingenieurs. Herausgeber: Sauer/Szabó. 1. Teil. DM 88,—; US $ 24.20
140. Mathematische Hilfsmittel des Ingenieurs. Herausgeber: Sauer/Szabó. 2. Teil. DM 136,—; US $ 37.40
141. Mathematische Hilfsmittel des Ingenieurs. Herausgeber: Sauer/Szabó. 3. Teil. DM 98,—; US $ 27.00
142. Mathematische Hilfsmittel des Ingenieurs. Herausgeber: Sauer/Szabó. 4. Teil. DM 124,—; US $ 34.10
143. Schur/Grunsky: Vorlesungen über Invariantentheorie. DM 32,—; US $ 8.80
144. Weil: Basic Number Theory. DM 48,—; US $ 12.00
145. Butzer/Berens: Semi-Groups of Operators and Approximation. DM 56,—; US $ 14.00
146. Treves: Locally Convex Spaces and Linear Differential Equations. DM 36,—; US $ 9.90
147. Lamotke: Semisimpliziale algebraische Topologie. DM 48,—; US $ 13.20
148. Chandrasekharan: Introduction to Analytic Number Theory. DM 28,—; US $ 7.00
149. Sario/Oikawa: Capacity Functions. DM 96,—; US $ 24.00
150. Iosifescu/Theodorescu: Random Processes and Learning. DM 68,—; US $ 18.70
151. Mandl: Analytical Treatment of One-dimensional Markov Processes. DM 36,—; US $ 9.80
152. Hewitt/Ross: Abstract Harmonic Analysis. Voll. II. Structure and Analysis for Compact Groups. Analysis on Locally Compact Abelian Groups. DM 140,—; US $ 38.50
153. Federer: Geometric Measure Theory. DM 118,—; US $ 29.50
154. Singer: Bases in Banach Spaces I. DM 112,—; US $ 30.80
155. Müller: Foundations of the Mathematical Theory of Electromagnetic Waves. DM 58,—; US $ 16.00
156. van der Waerden: Mathematical Statistics. DM 68,—; US $ 18.70
157. Prohorov/Rozanov: Probability Theory. DM 68,—; US $ 18.70
158. Constantinescu/Cornea: Potential Theory on Harmonic Spaces. In preparation
159. Köthe: Topological Vector Spaces I. DM 78,—; US $ 21.50
160. Agrest/Maksimov: Theory of Incomplete Cylindrical Functions and their Applications. In preparation
161. Bhatia/Szegö: Stability Theory of Dynamical Systems. DM 58,—; US $ 16.00
162. Nevanlinna: Analytic Functions. DM 76,—; US $ 21.00
163. Stoer/Witzgall: Convexity and Optimization in Finite Dimensions I. DM 54,—; US $ 14.90
164. Sario/Nakai: Classification Theory of Riemann Surfaces. DM 98,—; US $ 27.00
165. Mitrinović: Analytic Inequalities. DM 88,—; US $ 26.00

166. Grothendieck/Dieudonné: Eléments de Géométrie Algébrique I. DM 84,—; US $ 23.10
167. Chandrasekharan: Arithmetical Functions. DM 58,—; US $ 16.00
168. Palamodov: Linear Differential Operators with Constant Coefficients. DM 98,—; US $ 27.00
169. Rademacher: Topics in Analytic Number Theory. In preparation
170. Lions: Optimal Control of Systems Governed by Partial Differential Equations. DM 78,—; US $ 21.50
171. Singer: Best Approximation in Normed Linear Spaces by Elements of Linear Subspaces. DM 60,—; US $ 16.50
172. Bühlmann: Mathematical Methods in Risk Theory. DM 52,—; US $ 14.30
173. F. Maeda/S. Maeda: Theory of Symmetric Lattices. DM 48,—; US $ 13.20
174. Stiefel/Scheifele: Linear and Regular Celestial Mechanics. In preparation
175. Larsen: An Introduction to the Theory of Multipliers. In preparation
176. Grauert/Remmert: Analytische Stellenalgebren. In Vorbereitung
177. Flügge: Practical Quantum Mechanics (I. Teil: Kapitel I + II). In preparation
178. Flügge: Practical Quantum Mechanics (II Teil: Kapitel III—VII). In preparation